Lecture Notes in Electrical Engineering 284

For further volumes:
http://www.springer.com/series/7818

Natarajan Meghanathan • Dhinaharan Nagamalai
Sanguthevar Rajasekaran
Editors

Networks and Communications (NetCom2013)

Proceedings of the Fifth International Conference on Networks & Communications

Springer

Editors
Natarajan Meghanathan
Department of Computer Science
Jackson State University
Jackson, MS, USA

Dhinaharan Nagamalai
Wireilla Net Solutions PTY Ltd
Melbourne, VIC, Australia

Sanguthevar Rajasekaran
Department of CSE
University of Connecticut
Storrs, CT, USA

ISSN 1876-1100
ISBN 978-3-319-03691-5
DOI 10.1007/978-3-319-03692-2
Springer Cham Heidelberg New York Dordrecht London

ISSN 1876-1119 (electronic)
ISBN 978-3-319-03692-2 (eBook)

Library of Congress Control Number: 2013958445

© Springer International Publishing Switzerland 2014
This work is subject to copyright. All rights are reserved by the Publisher, whether the whole or part of the material is concerned, specifically the rights of translation, reprinting, reuse of illustrations, recitation, broadcasting, reproduction on microfilms or in any other physical way, and transmission or information storage and retrieval, electronic adaptation, computer software, or by similar or dissimilar methodology now known or hereafter developed. Exempted from this legal reservation are brief excerpts in connection with reviews or scholarly analysis or material supplied specifically for the purpose of being entered and executed on a computer system, for exclusive use by the purchaser of the work. Duplication of this publication or parts thereof is permitted only under the provisions of the Copyright Law of the Publisher's location, in its current version, and permission for use must always be obtained from Springer. Permissions for use may be obtained through RightsLink at the Copyright Clearance Center. Violations are liable to prosecution under the respective Copyright Law.
The use of general descriptive names, registered names, trademarks, service marks, etc. in this publication does not imply, even in the absence of a specific statement, that such names are exempt from the relevant protective laws and regulations and therefore free for general use.
While the advice and information in this book are believed to be true and accurate at the date of publication, neither the authors nor the editors nor the publisher can accept any legal responsibility for any errors or omissions that may be made. The publisher makes no warranty, express or implied, with respect to the material contained herein.

Printed on acid-free paper

Springer is part of Springer Science+Business Media (www.springer.com)

Preface

The Fifth International Conference on Networks & Communications (NETCOM-2013) was held in Chennai, India, from December 27–28, 2013. NETCOM attracted many local and international delegates, presenting a balanced mixture of intellect from the East and from the West. The goal of this conference series is to bring together researchers and practitioners from academia and industry and share cutting-edge development in the field. The conference provided an excellent international forum for sharing knowledge and results in theory, methodology and applications of Computer Networks and Data Communications. Authors were invited to contribute to the conference by submitting articles that illustrate research results, projects, survey work and industrial experiences describing significant advances in areas focusing on computer networks, network protocols, wireless networks, data communication technologies and network security.

The NETCOM-2013 committees rigorously invited submissions for many months from researchers, scientists, engineers, students and practitioners related to the relevant themes and tracks of the conference. This effort guaranteed submissions from an unparalleled number of internationally recognized top-level researchers. All the submissions underwent a strenuous peer-review process which comprised expert reviewers. These reviewers were selected from a talented pool of Technical Committee members and external reviewers on the basis of their expertise. The papers were then reviewed based on their contributions, technical content, originality and clarity. The entire process, which includes the submission, review and acceptance processes, was done electronically. Extended versions of selected papers from the conference will be invited for publication in several international journals. All these efforts undertaken by the Organizing and Technical Committees led to an exciting, rich and a high quality technical conference program, which featured high-impact presentations for all attendees to enjoy, appreciate and expand their expertise in the latest developments in various research areas of Computer Networks and Data Communications. In closing, NETCOM-2013 brought together researchers, scientists, engineers, students and practitioners to exchange and share their experiences, new ideas and research results in all aspects of the main workshop themes and tracks, and to discuss the practical challenges encountered and the solutions adopted. We would like to thank the General and Program Chairs, organization staff, the members of the Technical Program Committees and external reviewers for their excellent and tireless work. We sincerely wish that all attendees benefited scientifically from the conference and wish them every success in their research.

It is the humble wish of the conference organizers that the professional dialogue among the researchers, scientists, engineers, students and educators continues beyond the event and that the friendships and collaborations forged will linger and prosper for many years to come.

Jackson, MS, USA	Natarajan Meghanathan
Melbourne, VIC, Australia	Dhinaharan Nagamalai
Storrs, CT, USA	Sanguthevar Rajasekaran

Organization

General Chairs

- **Sanguthevar Rajasekaran**, University of Connecticut, USA
- **Jagannathan Sarangapani**, Missouri University of Science and Technology, USA

Steering Committee

- **Brajesh Kumar Kaushik**, Indian Institute of Technology, Roorkee, India
- **Natarajan Meghanathan**, Jackson State University, USA
- **Khoa N. Le**, University of Western Sydney, Australia
- **Rakhesh Singh Kshetrimayum**, Indian Institute of Technology, Guwahati, India
- **Dhinaharan Nagamalai**, Wireilla Net Solutions PTY Ltd, Australia
- **Salah M. Saleh Al-Majeed**, University of Essex, United Kingdom

Program Committee Members

- **Jacques Demerjian**, Communications & Systems, France
- **Jan Zizka**, SoNet/DI, FBE, Mendel University in Brno, Czech Republic
- **Sarmistha Neogy**, Jadavpur University, India
- **Hwangjun Song**, POSTECH (Pohang University of Science and Technology), South Korea
- **Rakhesh Singh Kshetrimayum**, Indian Institute of Technology, Guwahati, India
- **Krzysztof Walkowiak**, Wroclaw University of Technology
- **Ioannis Karamitsos**, University of Aegean, Greece
- **Ramayah**, Universiti Sains Malaysia, Malaysia
- **Khoa N. Le**, Griffith School of Engineering, Queensland, Australia

- **T.G. Basavaraju**, National Institute of Technology Karnataka (NITK), India
- **Solange Rito Lima**, University of Minho, Portugal
- **Sherif S. Rashad**, Morehead State University, USA
- **Dhinaharan Nagamalai**, Wireilla Net Solutions PTY Ltd, Australia
- **David C. Wyld**, Southeastern Louisiana University, USA
- **Selma Boumerdassi**, CNAM/CEDRIC, France
- **H.V. Ramakrishnan**, Bharath University, India
- **Sattar B. Sadkhan**, University of Babylon, Iraq
- **Eric Renault**, Institut Telecom – Telecom SudParis, Evry, France
- **Alvin Lim**, Auburn University, USA
- **Debasis Giri**, Haldia Institute of Technology, India
- **S. Li**, Swansea University, UK
- **Rushed Kanawati**, LIPN – University Paris 13, France
- **Cristina Ribeiro**, University of Waterloo, Canada
- **Samiran Chattopadhyay**, Jadavpur University, India
- **Rajarshi Roy**, IIT, Kharagpur, India
- **S.A.V. Satyamurty**, Indira Gandhi Centre for Atomic Research, India
- **Laiali Almazaydeh**, University of Bridgeport, USA
- **Shrikant K. Bodhe**, Bosh Technologies, India
- **Alireza mahini**, Islamic Azad University, Gorgan, Iran
- **N.K. Choudhari**, Smt. Bhagwati Chaturvedi College of Engineering, India
- **Christos Politis**, Kingston University, UK
- **Ayman Khalil**, Institute of Electronics and Telecommunications of Rennes, France
- **Sridharan**, CEG Campus – Anna University, India
- **Mohamed Hassan**, American University of Sharjah, UAE
- **Zuqing Zhu**, Cisco Systems, USA
- **Quan (Alex) Yuan**, University of Wisconsin-Stevens Point, USA
- **Henrique J.A. Holanda**, UERN – Universidade do Estado do Rio Grande do Norte
- **Ajay K. Sharma**, Dr. B R Ambedkar National Institute of Technology, India
- **Shrirang Ambaji Kulkarni**, National Institute of Engineering, India
- **Shin-ichi Kuribayashi**, Seikei University, Japan
- **Abdel Salhi**, University of Essex, United Kingdom
- **Antonio Liotta**, Eindhoven University of Technology, the Netherlands
- **Emmanuel Jammeh**, University of Plymouth, United Kingdom
- **Ghaida Al-Suhail**, Basrah University, Iraq
- **Hao-En Chueh**, Yuanpei University, Taiwan, Republic of China
- **John Woods**, University of Essex, United Kingdom
- **Ken Guild**, University of Essex, United Kingdom
- **Martin Fleury**, University of Essex, United Kingdom
- **Mohammad M. Banat**, Jordan University of Science and Technology, Jordan
- **Nadia Qadri**, University of Essex, United Kingdom
- **S. Hariharan**, J.J. College of Engineering, India
- **Yasir Qadri**, University of Essex, United Kingdom

- **Wichian Sittiprapaporn**, Mahasarakham University, Thailand
- **Mahi Lohi**, University of Westminster, UK
- **Houcine Hassan**, Univeridad Politecnica de Valencia, Spain
- **Mohammed Ghanbari**, University of Essex, United Kingdom
- **Abdulrahman Yarali**, Murray State University, USA
- **Asmaa Shaker Ashoor**, Babylon University, Iraq
- **Chin-Chih Chang**, Chung Hua University, Taiwan
- **Doina Bein**, The Pennsylvania State University, USA
- **Hossein Jadidoleslamy**, University of Zabol, Zabol, Iran
- **Kayhan Erciyes**, Izmir University, Turkey
- **M. Mohamed Ashik**, Salalah College of Technology, Oman
- **Mohamed Fahad AlAjmi**, King Saud University, Saudi Arabia
- **Moses Ekpenyong**, University of Edinburgh, Nigeria
- **Natarajan Meghanathan**, Jackson State University, USA
- **Nazmus Saquib**, University of Manitoba, Canada
- **Ruchi Tuli**, Yanbu University College, Kingdom of Saudi Arabia
- **Serguei A. Mokhov**, Concordia University, Canada
- **V. Sundarapandian**, Vel Tech Dr. RR & Dr. SR Technical University, India
- **Pinaki Sarkar**, Jadavpur University, India
- **S. Taruna**, Banasthali University, India
- **S. Rajaram**, Thiagarajar College of Engineering, India
- **Uday Nuli**, Textile and Engineering Institute, Ichalkaranji, India
- **Shun Hattori**, Muroran Institute of Technology, Japan
- **Yoram Haddad**, Jerusalem College of Technology/Ben Gurion University, Israel
- **Cathryn Peoples**, University of Ulster, United Kingdom
- **Antonio Ruiz-Martinez**, University of Murcia, Spain
- **Paulo R.L. Gondim**, University of Brasilia, Brazil
- **Josip Lorincz**, University of Split, Croatia
- **Jose-Fernan Martinez-Ortega**, Universidad Politecnica de Madrid – UPM, Spain
- **Noor Zaman**, College of Computer Science & IT, King Faisal University, Saudi Arabia
- **Hangwei**, Western Reserve University, USA
- **Nuno M. Garcia**, Universidade Lusofona de Humanidades e Tecnologias, Portugal
- **Rachida Dssouli**, Concordia University, Canada
- **Jaime Lloret**, Polytechnic University of Valencia, Spain
- **Daqiang Zhang**, Nanjing Normal University, China
- **Juan Jose Martinez Castillo**, Ayacucho University, Venezuela
- **Malamati Louta**, University of Western Macedonia, Greece
- **Malka N. Halgamuge**, The University of Melbourne, Australia
- **Jose Neuman de Souza**, Federal University of Ceara, Brazil
- **Iwan Adhicandra**, University of Pisa, Italy
- **Bob Natale**, MITRE, USA

- **Hamza Aldabbas**, De Montfort University, UK
- **Behnam Dezfouli**, University Technology Malaysia (UTM), Malaysia
- **Ehsan Heidari**, Islamic Azad University Doroud Branch, Iran
- **Jadidoleslamy**, University of Zabol, Iran
- **M. Nadeem Baig**, Riyadh, Kingdom of Saudi Arabia
- **Nisar Hundewale**, University of Maryland University College, USA
- **Omar Almomani**, Jadara University, Jordan
- **Paulo Martins Maciel**, Federal University of Pernambuco, Brazil
- **Phan Cong Vinh**, NTT University, Vietnam
- **Raed Alsaqour**, Universiti Kebangsaan Malaysia, Malaysia
- **Sajid Hussain**, Fisk University, USA
- **Sherimon P.C.**, Arab Open University, Sultanate of Oman
- **Somayeh Mohammadi**, Islamic Azad University, Iran
- **Mahdi Aiash**, Middlesex University, UK
- **Almir Pereira Guimaraes**, Federal University of Alagoas, Brazil
- **Ahmed Ezz Eldin Khaled**, Cairo University, Egypt
- **Subhankar Mishra**, University of Florida, USA
- **Gary Campbell**, University College of the Caribbean, Jamaica
- **Hazem M. Al-Najjar**, Taibah University, Kingdom of Saudi Arabia
- **Yingchi Mao**, Hohai University, China
- **Rabie Ramadan**, Cairo University, Egypt
- **Behnam Dezfouli**, Universiti Teknologi Malaysia (UTM), Malaysia
- **Parminder S. Reel**, The Open University, United Kingdom
- **Suleyman Kondakci**, Izmir University of Economics, Turkey
- **XU Mengxi**, Nanjing Institute of Technology, China

Committee Members/Reviewers

- **Abd El-Aziz Ahmed**, Anna University, Chennai, India
- **Alagu S.**, Alagappa University, India
- **Anita Seth**, DAVV University, India
- **Ankit Gandhi**, HCL Technologies Ltd, Bangalore, India
- **Annapurna P. Patil**, M S Ramaiah Institute of Technology, India
- **Arun Vijayaragavan**, RMD Engineering College, India
- **Ashutosh Kumar Dubey**, Trinity Institute of Technology & Research, Bhopal, India
- **Bhupendra Suman**, IIT Roorkee, India
- **Bikesh Kumar Singh**, National Institute of Technology Raipur, India
- **Binod Kumar Pattanayak**, Siksha 'O' Anusandhan University, India
- **D. Aruna Kumari**, K L University, India
- **D.B. Bhoyar**, Yeshwantrao Chavan College of Engineering, India
- **Daniel A.K.**, M.M.M. Engineering College, Gorakhpur, India
- **Deepa N.**, VIT University, Vellore, India
- **Demian D'Mello**, St. Joseph Engineering College, Mangalore, India

- **Devesh Jinwala**, S V National Institute of Technology, India
- **Diptoneel Kayal**, West Bengal University of Technology, India
- **E. George Dharma Prakash Raj**, Bharathidasan University, India
- **Ferdous Ahmed Barbhuiya**, IIT, Guwahati, India
- **G. Suresh Babu**, CBIT Hyderabad, India
- **Hameem Shanavas**, MVJ College of Engineering, India
- **Hemanta Kumar Kalita**, North-Eastern Hill University (NEHU), India
- **Jagadeesh B.**, G.V.P. College of Engineering, India
- **Jitendra Singh Jadon**, Amity University, India
- **Kahkashan Tabassum**, Maulana Azad National Urdu University, India
- **Keshavamurthy B.N.**, Manipal Institute of Technology, India
- **Kilari VeeraSwamy**, QIS College of Engineering and Technology, India
- **Kiran Kumari Patil**, Reva ITM Bangalore, India
- **Kumar Abhishek**, National Institute of Technology, Patna, India
- **M.K. Kavitha Devi**, Thiagarajar College of Engineering, India
- **M. Sushanth Babu**, Vaagdevi College of Engineering, India
- **M. Upendra Kumar**, Mahatma Gandhi Institute of Technology, India
- **Madhumita Chatterjee**, Pillai's Institute of Information Technology, India
- **Mahabaleshwar S.K.**, Basaveshwar Engineering College, India
- **Mamatha Balachandra**, Manipal Institute of Technology, India
- **Manish Mahajan**, Chandigarh Engineering College, India
- **Manjaiah D.H.**, Mangalore University, India
- **Mansaf Alam**, Jamia Millia Islamia, New Delhi, India
- **Manu Sood**, Himachal Pradesh University, India
- **Mayank Kumar Goyal**, Sharda University, India
- **Md. Amir Khusru Akhtar**, Cambridge Institute of Technology, India
- **M.H.M. Krishna Prasad**, JNTUK University College of Engineering, India
- **Musheer Ahmad**, Jamia Millia Islamia New Delhi, India
- **Mydhili K. Nair**, MSRIT Bangalore, India
- **Nagaraju A.**, Vaagdevi College of Engineering, India
- **Neeraj kumar**, Thapar University, India
- **Nishant Doshi**, S V National Institute of Technology, India
- **Nityananda Sarma**, Tezpur University, India
- **P.E.S.N. Krishna Prasad**, Prasad V. Potluri Siddhartha Institute of Technology, India
- **Pankaj Gupta**, Capillary Technologies, India
- **Parminder Singh**, Chandigarh Engineering College, India
- **Praveen Goyal**, Shiv Kumar Singh Institute of Technology & Science, India
- **Pushpendra Pateriya**, Lovely Professional University, India
- **R. Deepa**, Amrita Vishwa Vidhyapeetham, India
- **Ram Mangrulkar**, B.D. College of Engineering, India
- **Rama Krishna C.**, NITTTR Chandigarh, India
- **Ramesh Babu H.S.**, Acharya Institute of Technology, India
- **Rangisetty Nirmala Devi**, KITS Warangal, India
- **Ravendra Singh**, MJP Rohilkhand University, India

- **Ravinder Ahuja**, Jaypee University of Information Technology, India
- **S. Geetha**, Thiagarajar College of Engineering, India
- **S. Hariharan**, TRP Engineering College, India
- **S.P. Balakannan**, Kalasalingam University, India
- **Samarendra Nath Sur**, Sikkim Manipal Institute of Technology, India
- **Sandeep M. Chaware**, Bivrabai Sawant College of Engineering & Research, India
- **Sandeep Singh**, Babasaheb Bhimrao Ambedkar University, India
- **Sanjay Singh**, Manipal Institute of Technology, India
- **Sankhayan Choudhury**, University of Calcutta, India
- **Santosh Biswas**, IIT, Guwahati, India
- **Sesha Bhargavi Velagaleti**, JNTUH, Hyderabad, India
- **Shivani Mishra**, MNNIT, Allahabad, India
- **Shobha K.R.**, M S Ramaiah Institute of Technology, India
- **Shriram K.V.**, Amrita University, India
- **Soumen Kanrar**, Vehere Interactive Pvt Ltd, Calcutta, India
- **Sudarshan Patel**, A.D. Patel Institute of Technology, India
- **Sumthy S.**, VIT University, Vellore, India
- **Suparna DasGupta**, JIS College of Engineering, India
- **Sushil Kumar**, Jawaharlal Nehru University, India
- **Sutanu Ghosh**, West Bengal University of Technology, India
- **Syed Imtiyaz Hassan**, Jamia Hamdard, New Delhi, India
- **T. Ramkumar**, A.V.C. College of Engineering, India
- **Tapan Jain**, Jaypee University of Information Technology, India
- **Usha Sharma**, Banasthali University, India
- **V.K. Pachghare**, College of Engineering, Pune, India
- **V. Shanthi**, St. Joseph's College of Engineering, India
- **V.V. Rama Prasad**, Sree Vidyanikethan Engineering College, India
- **Zeenat Rehena**, Aliah University, India
- **Amol Vasudeva**, Jaypee University of Information Technology, India

Contents

Part I Measurement and Performance Analysis

1 An Evaluation of EpiChord in OverSim 3
  Jamie Furness, Farida Chowdhury, and Mario Kolberg

2 Enhanced Back-Off Technique for IEEE 802.15.4
  WSN Standard 21
  Aditi Vutukuri, Saayan Bhattacharya, Tushar Raj,
  Sridhar, and Geetha V

3 Efficient Retransmission QoS-Aware MAC Scheme in Wireless
  Sensor Networks 31
  M. Kumaraswamy, K. Shaila, V. Tejaswi, K.R. Venugopal,
  S.S. Iyengar, and L.M. Patnaik

4 Performance Analysis of Correlated Channel
  for UWB-MIMO System 43
  Mihir Narayan Mohanty, Monalisa Bhol, and Sanjat Kumar Mishra

Part II Network Architectures, Protocols and Routing

5 Node Failure Time Analysis for Maximum Stability Versus
  Minimum Distance Spanning Tree Based Data Gathering
  in Mobile Sensor Networks 55
  Natarajan Meghanathan

6 Wireless Networks: An Instance of Tandem
  Discrete-Time Queues 69
  Nikhil Singh and Ramavarapu S. Sreenivas

7 An Equal Share Ant Colony Optimization Algorithm
  for Job Shop Scheduling Adapted to Cloud Environments 81
  Rajesh Chaukwale and Sowmya Kamath S

8   A Dynamic Trust Computation Model for Peer
    to Peer Network 93
    Tarek Helmy

9   Road Traffic Management Using Egyptian Vulture
    Optimization Algorithm: A New Graph Agent-Based
    Optimization Meta-Heuristic Algorithm 107
    Chiranjib Sur and Anupam Shukla

Part III   Ad Hoc and Sensor Networks

10 A Novel Bloom Filter Based Variant of Elliptic Curve Digital
    Signature Algorithm for Wireless Sensor Networks 125
    Vivaksha Jariwala, Prafulla Kumar, and Devesh C. Jinwala

11 Energy Efficient Localization in Wireless Sensor Networks 139
    A.V. Sutagundar, S.S. Shirabur, and V.S. Bennur

Part IV   Network Security, Trust and Privacy

12 Data Integrity Verification in Hybrid Cloud Using TTPA 149
    T. Subha and S. Jayashri

13 A Comparative Study of Data Perturbation Using Fuzzy
    Logic to Preserve Privacy 161
    Thanveer Jahan, G. Narasimha, and C.V. Guru Rao

14 Vector Quantization in Language Independent Speaker
    Identification Using Mel-Frequency Cepstrum Co-efficient 171
    D. Ambika and V. Radha

15 Improved Technique for Data Confidentiality
    in Cloud Environment 183
    Pratyush Ranjan, Preeti Mishra, Jaiveer Singh Rawat,
    Emmanuel S. Pilli, and R.C. Joshi

16 A Hybrid-Based Feature Selection Approach for IDS 195
    Amrita and P. Ahmed

17 Reckoning Minutiae Points with RNA-FINNT Augments
    Trust and Privacy of Legitimate User and Ensures Network
    Security in the Public Network 213
    Kuljeet Kaur and G. Geetha

18 Empirical Study of Email Security Threats
    and Countermeasures 229
    Dhinaharan Nagamalai, Beatrice Cynthia Dhinakaran,
    Abdulkadir Ozcan, Ali Okatan, and Jae-Kwang Lee

19 Improving Business Intelligence Based on Frequent
    Itemsets Using k-Means Clustering Algorithm 243
    Prabhu Paulraj and Anbazhagan Neelamegam

20 Reputation-Based Trust Management for Distributed
    Spectrum Sensing 255
    Seamus Mc Gonigle, Qian Wang, Meng Wang,
    Adam Taylor, and Eamonn O. Nuallain

21 Ontology Based Multi-Agent Intrusion Detection System
    for Web Service Attacks Using Self Learning 265
    Krupa Brahmkstri, Devasia Thomas, S.T. Sawant,
    Avdhoot Jadhav, and D.D. Kshirsagar

22 A Novel Key Exchange Protocol Provably Secure
    Against Man-in-the-Middle Attack 275
    Abhijit Chowdhury, Shubhajit Nath, and Jaydeep Howlader

23 Neutralizing DoS Attacks on Linux Servers 281
    G. Rama Koteswara Rao and A. Pathanjali Sastri

Part V Wireless and Mobile Networks

24 Cache Coherency Algorithm to Optimize Bandwidth
    in Mobile Networks 297
    Abhinandan Ramaprasath, Karthik Hariharan, and Anand Srinivasan

25 Connected Dominating Set Based Scheduling
    for Publish-Subscribe Scenarios in WSN 307
    P. Kaviya and R. Ramalakshmi

26 Ubiquitous Compaction Monitoring Interface for Soil
    Compactor: A Web Based Approach 319
    R. Prakash, K. Suresh, S. Mydhile Shanmugam, and C.P. Koushik

Author Index 331

Subject Index 333

Part I
Measurement and Performance Analysis

Chapter 1
An Evaluation of EpiChord in OverSim

Jamie Furness, Farida Chowdhury, and Mario Kolberg

Abstract EpiChord is a Distributed Hash Table (DHT) algorithm which supports data storage/retrieval in large scale distributed systems. It removes the typical $O(logn)$-state-per-node restriction imposed by the majority of other DHT topologies by employing a reactive routing state maintenance strategy that amortizes network maintenance costs into lookup queries. Under ideal condition, EpiChord's lookup performance can approach $O(1)$ hops – with maintenance costs comparable to traditional multi-hop DHTs. This paper presents an implementation of EpiChord in OverSim, and validates the performance of our model against the performance reported in the original EpiChord paper. We also present some adjustments to the algorithm to remove a discrepancy and then compare our modified results with the original ones. Finally, we present additional results showing the EpiChord algorithm is stable over time and performs well for larger networks.

1.1 Introduction

Distributed Hash Tables (DHTs) [1] supported Peer-to-Peer (P2P) applications are an ideal substrate for building large scale distributed systems because they are self-organizing, adaptable and scalable and offer efficient routing between nodes within a bounded number of hops. EpiChord [2] is a DHT lookup algorithm which demonstrates that node state restrictions can be relaxed which were imposed by the majority of other DHT algorithms by using a reactive routing state maintenance strategy. Nodes piggyback additional network information on lookup queries to keep their routing state up-to-date. This makes EpiChord ideally suited to large scale environments. This paper discusses an implementation [3] of EpiChord within

J. Furness • F. Chowdhury (✉) • M. Kolberg
Computing Science and Mathematics, University of Stirling, Stirling, Scotland
e-mail: jrf@cs.stir.ac.uk; fch@cs.stir.ac.uk; mko@cs.stir.ac.uk

the OverSim Simulator [4]. The model is validated against the original EpiChord paper. Specifically, the contributions of the paper are as follows:

- An independent evaluation of EpiChord, by comparing results from our simulation model to the results presented in the original EpiChord paper.
- Performance evaluation in multiple scenarios, defined in the original paper, which test both the routing and maintenance algorithms of the model.
- Amendments to the original model together with a comparison of the results obtained from our model against the corrected results from the original model.
- Performance evaluation of EpiChord in larger networks and for longer simulations.
- A freely available EpiChord model in OverSim.
- A review of available simulators.

The original implementation of EpiChord was a model for the SSFNet simulation framework [5] which is not publicly available. The authors are not aware of other EpiChord models which are publicly available. In other work [6] we have validated the models for both Chord and Pastry in OverSim.

The remainder of the paper is structured as follows: Sect. 1.2 discusses related work, Sect. 1.3 provides an overview of the EpiChord DHT algorithm, Sect. 1.4 compares network simulators, Sect. 1.5 discusses implementation details of the EpiChord model in OverSim, Sect. 1.6 presents an evaluation of results after changes to the original EpiChord model. Section 1.7 presents validating results from our EpiChord model as well as results demonstrating EpiChord's scalability. Section 1.8 concludes this paper.

1.2 Related Work

A large number of multi-hop structured Peer-to-Peer (P2P) algorithms have been proposed [1]. These algorithms are characterized by O(log N) hop count. Because each overlay hop translates to potentially many hops in the underlying network, multi-hop overlays have a relatively poor latency characteristic for connecting large numbers of peers. Consequently, systems have been developed to trade-off latency for larger routing tables. However these designs lead to increased network traffic for managing the larger routing tables. Thus efficient overlay maintenance in O(1)-hop (one-hop) overlays is an important research question. Two techniques have emerged [1] for maintaining routing tables in overlays: active stabilization where peers have fixed communication to maintain a target routing table accuracy, and opportunistic updating where routing table maintenance depends on lookup load and available bandwidth.

An example of an active stabilization algorithm is EDRA (Event Detection and Reporting Algorithm) used in the D1HT one-hop overlay [7]. EDRA has been proposed to give reasonable message rate for high levels of routing table accuracy. For example, D1HT has up to an order magnitude lower maintenance bandwidth usage compared to the OneHop [8], another active stabilization one-hop overlay.

EDRA* [9] offers some improvements over EDRA. Examples of opportunistic overlay maintenance include EpiChord [2] (used in this paper) and Accordion [10].

Kelips [11] is a O(1)-hop overlay which uses an epidemic multicast protocol for exchanging overlay membership and other soft state between nodes. Such a protocol consists of two sub-protocols: a multicast data dissemination protocol and a gossip protocol to exchange message history for reliability purposes.

Accordion [10] is a variable hop overlay, in which a peer limits its routing table update message level based on its available bandwidth. During periods of low bandwidth, routing table accuracy can approach that of multi-hop overlays while for higher bandwidth, routing table accuracy reaches one-hop. Accordion uses recursive parallel lookups so as to maintain fresh routing table entries in its neighborhood of the overlay and reduce the probability of timeout. Note that recursive parallel lookups create more load on the target peer compared to iterative parallel lookups.

1.3 EpiChord Background

EpiChord [2] is a DHT algorithm which can achieve one-hop lookup performance under lookup intensive workloads, and at worst case $O(log_2(N))$ hop, as offered in many multi-hop networks. As the name suggests, EpiChord is based on the Chord DHT [12]. Like Chord, EpiChord is organized in a one-dimensional circular address space where each node is assigned a unique node identifier. The node responsible for a key is the node whose identifier most closely follows the key. In addition to maintaining a list of k succeeding nodes, EpiChord also maintains a list of the k preceding nodes. Instead of maintaining a finger table, as in Chord, EpiChord maintains a cache of nodes. Nodes update their cache by observing lookup traffic, and add an entry anytime they learn of a node not already in the cache. Nodes in the cache each have a timeout, resulting in stale nodes being removed.

In general terms EpiChord can be thought of as Chord with a cache of extra node addresses. As such the routing algorithm is similar to that in Chord. With a well populated cache this results in lookup performance of one hop. Under high churn the performance drops to that of Chord, $O(log_2(N))$ hops in the worst case.

1.3.1 Lookup Algorithm

EpiChord uses an iterative lookup algorithm, as it avoids sending redundant queries when using parallel requests. It also allows the querying node to receive all information related to the query path, and hence updates its cache with new entries. To lookup a data item with the key id, a node will initiate p queries in parallel – to the node immediate succeeding id and to the $p-1$ nodes preceding id. When queried, a node will respond as follows (l and p are both system parameters):

- If it owns id, it will return the value associated with id, and information on its predecessor and successor.

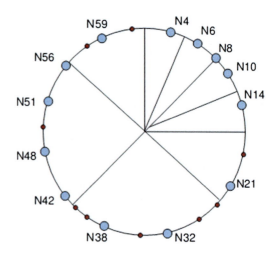

Fig. 1.1 Example of slicing of address space for N8

- If it is a predecessor of *id* relative to the querying node, it will provide information about its successor and the *l* best next hops towards the destination.
- If it is a successor of *id* relative to the querying node, it will provide information on its predecessor and the *l* best next hops towards the destination.

When a reply is received, further queries are dispatched in parallel if the querying node learns about any node closer to the target *id* than the best successor and predecessor nodes that have already responded.

1.3.2 Cache Invariant

To guarantee worst case lookup performance of $O(log_2(N))$ each node divides the address space into two sets of exponentially smaller slices, as seen in Fig. 1.1. Each node maintains their cache such that every slice contains at least $\frac{j}{1-\gamma}$ cache entries at all times, where j is a network parameter and γ is a local estimate of the probability that a cache entry is out-of-date. Nodes periodically check their cache slices to ensure that there are sufficient unexpired cache entries. To calculate γ, each node keeps track of n_p, the number of messages sent, and n_t, the number of messages which timed out. γ is calculated using n_t / n_p. In addition, n_p and n_t are periodically (when the cache is flushed) multiplied by a network parameter δ to obtain exponentially weighted moving averages.

1.3.3 Routing Table Updates

Each node periodically probes their immediate neighbours to ensure that they are still alive. The delay between these stabilization attempts is calculated based on the observed lifetime of nodes in the finger cache. For this reason the finger cache also contains a map of dead nodes, and the observed lifetime is calculated by taking the time between first learning of the node (s_{start}) until learning of its death (s_{end}). The observed lifetime for each dead node is averaged, and the obtained estimate is then multiplied by the lifetime estimate multiplier, ω, to calculate when the next stabilization attempt should be scheduled.

$$s = \frac{\sum s_{end} - s_{start}}{n} \cdot \omega \qquad (1.1)$$

In case where the sample size, n, is less than 5, the stabilization interval is simply set to the network parameter s.

With active propagation, nodes will inform their neighbours of any detected changes in the successor or predecessor lists as soon as they happen, rather than waiting for the next stabilization attempt. This increases the maintenance bandwidth when under high churn, however also results in more accurate successor and predecessor lists, and hence fewer false-negatives.

If a node has an outdated view of the local key space that they are responsible for, they may fail to respond correctly to all queries. By including their believed predecessor and successor in the query response, the querying node can either make a step towards the destination or, if the believed predecessor does not respond, determine that the responsible node is dead. This false-negative detection allows the querying node to resolve the lookup correctly. If a false-negative is detected, the querying node will immediately inform the new responsible node that their predecessor has failed and now they should be responsible for the requested key.

1.4 Review of Simulators

Before deciding on OverSim, a detailed review of other available and active P2P network simulators was carried out. A summary of these tools is provided in Table 1.1.

PeerSim [13] is written in Java. Its main focus is to provide high scalability and can handle a network of up to 10^6 nodes. However, this scalability comes at the cost of not including a model of the behavior of the underlying communication network, e.g. TCP/IP stack and latencies.

P2PSim [14] is a discrete event simulator for P2P overlays written in C++. It supports Chord, Accordion, Koorde, Kelips, Tapestry, and Kademlia. However, these implementations are specific to P2PSim and do not model all features of the protocols. P2PSim has been simulated with up to 3,000 nodes using the Chord implementation. This simulator is largely undocumented and therefore hard to extend.

Overlay Weaver [15] is a toolkit for P2P Overlays written in Java. It has been tested with tens of thousands of nodes (their website quotes 300,000). Chord, Kademlia, Pastry, Tapestry and Koorde are available. The simulations have to be run in real-time environments and there is no statistical output which makes its use very limited.

Table 1.1 A comparison of available active P2P simulators

Simulator	P2P protocols	Network size	Language
PeerSim	Collection of internally developed P2P models	$>10^6$	Java
P2PSim	Chord, Accordion, Koorde, Kelips, Tapestry, Kademlia	3,000	C++
Overlay weaver	Chord, Kademlia, Koorde, Pastry, Tapestry and FRT-Chord	Tens of thousand	Java
PlanetSim	Chord, Symphony	100,000	Java
NS2	Gnutella	N/A	C++/OTcl
SSFNet	Chord, EpiChord	33,000	Java/C++/DML
OverSim	Chord, Kademlia, Pastry, Bamboo, Broose, Gia	100,000	C++
PeerfactSim.Kom	CAN, Chord, Kademlia, Gia, C-DHT, Gnutella 0.4/0.6, Pastry	50,000	Java
D-P2P-Sim+	Chord	400,000	Java

PlanetSim [16] is a discrete event simulation framework for both structured and unstructured overlays, written in Java. It has a modular, well-structured architecture and services can be re-used for other overlays. Chord and Symphony models exist and can consist of up to 100,000 nodes. However, it provides rather limited support to collect statistics. It also has a very simplified underlying network layer without any consideration of bandwidth and latency costs.

NS2 [17] is a discrete-event simulator that provides substantial support for simulation of lower layer protocols. Only one P2P protocol, Gnutella, is available in NS2. Simulations in NS2 are constructed using C++ and OTcl. It is mostly used for small networks and is generally unsuitable for large scale P2P overlay networks.

SSFNet [5] is a discrete-event simulation framework written in Java and C++. This framework is built on the Scalable Simulation Framework (SSF) and uses the Domain Modeling Language (DML) to configure networks. Chord and EpiChord have been implemented in SSFNet. There is a claim that SSFNet manages to run models with 33,000 nodes, however, the authors of the original EpiChord paper [2] and ourselves could not simulate networks with more than 10 k nodes.

OverSim [4] is an open-source P2P simulation framework for the OMNeT++ simulation environment. It provides a generic lookup mechanism and an RPC interface to facilitate additional protocol implementations. It allows large-scale simulations of simplified networks as well as complex heterogeneous underlay networks. Several P2P algorithms such as Chord, Kademlia, Bamboo, Broose, Koorde, NICE, NTree, Pastry, and GIA have been implemented in OverSim. Models can scale to over 100,000 nodes. More comprehensive surveys of P2P network simulators can be found in [18, 19].

PeerfactSim.Kom [20] is a discrete event based P2P simulator environment. Its focus is on being extendable and on large scale network models. This simulator offers the potential to model different types of peer-to-peer systems including distributed CDNs, streaming applications and overlay systems. It comes with a built-in churn generator. The simulator includes models of lower layers but does not yet include TCP.

D-P2P-Sim+ [21] is a distributed simulation environment which employs multithreading, asynchronous message passing and distributed environment with graphical user interface. There is little information on this simulator besides a short paper and poster. These report simulated network sizes of up to 400,000 nodes. It seems the only implemented overlay algorithm is Chord. However, the system is extendible and other algorithms could be implemented. Multiple computers running the simulator may be interconnected to achieve larger simulated network sizes.

Based on this study OverSim was selected for our experimentation due to its flexibility with respect to underlay characteristics and possible high scalability.

1.5 OverSim Implementation

OverSim [4] is designed as a modular simulation framework, with many common overlay features implemented as part of a generic base overlay class. OverSim provides message passing using Remote Procedure Calls (RPC), and supports both iterative and recursive routing. Applications within OverSim are split into multiple tiers, allowing an application to sit on-top of another application. These applications are implemented as modules and interface with overlays through the Key-Based Routing (KBR) API [22], which represents basic capabilities common to all structured overlays. As mentioned above, OverSim provides a number of different network models, for both structured and unstructured overlays. The OverSim architecture is illustrated in Fig. 1.2.

At the lower layer OverSim provides multiple underlay models to allow for inclusion of specific underlay characteristics in the simulation (at a cost of scalability), or underlay abstraction for increased scalability. Using the simple model, data packets are sent directly from one node to another by using a global routing table. The INET underlay model includes simulation models for all network layers. The single host underlay allows for simulation of a single node, connected to other OverSim instances over a real network.

Below we discuss some alterations which we made to the original EpiChord protocol when implementing it as an OverSim module.

1.5.1 Node Join Protocol

In the original EpiChord algorithm, upon receipt of a join request a node will instantly update their predecessor list and finger cache to include the joining node. In our implementation we found this was occasionally causing messages to be routed to nodes who are still in the process of joining, and not yet ready to correctly handle requests. To solve this issue we implemented a three-way handshake. In our implementation the joining node will send a final acknowledgment when they are ready to handle requests, indicating they can now be safely added as a predecessor.

Fig. 1.2 OverSim architecture

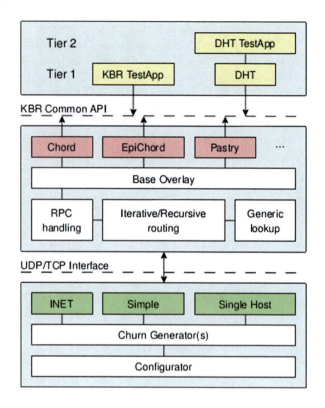

1.5.2 Lookup Algorithm

The OverSim framework provides modules for iterative and recursive routing, as can be seen in Fig. 1.2, with support for parallelism. While this makes implementation of many overlays easier and reduces duplicated code, only certain parts of the module can be easily overridden. This was a problem for EpiChord, primarily due to the non-linear order in which nodes are to be queried, and EpiChord's ability to check for false negative responses. To implement these features we had to make changes to the iterative routing module, allowing us to override additional parts of the module with code specific to EpiChord.

1.6 Results: Changes to the Original Model

1.6.1 Application Layer Lookups

In the original EpiChord model all lookup types (JOIN, MAINTENANCE, and APPLICATION) are included when calculating results. The KBRTestApp in OverSim only includes lookups it has initiated (APPLICATION) in the results. We feel this is actually a more useful metric for anyone wishing to build on-top of

1 An Evaluation of EpiChord in OverSim 11

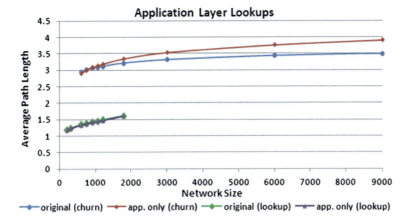

Fig. 1.3 Comparison of average path length with APPLICATION lookups only versus all lookup

EpiChord, so we instead recalculated the results from the original model using only APPLICATION lookups. A comparison of the average path lengths can be seen in Fig. 1.3; the other metrics remained unchanged.

In [23], authors proposed two generic classes of workloads: *lookup intensive* and *churn intensive*. These metrics were adopted by the EpiChord authors for experimentation. For the purposes of validating our model, we also adopt these two metrics. In the *lookup intensive* workload, node lifetimes are exponentially distributed with a mean of 10 min, and each node performs lookups on average every 0.5 s. In this scenario the background maintenance traffic is negligible compared to the active lookup rate. In the *churn intensive* workload, node lifetimes are again exponentially distributed with a mean of 10 min, however this time each node only performs lookups on average every 100 s. In this scenario the lookup rate is so low, most of the lookups captured are lookups arising from node joins and cache maintenance.

Figure 1.3 shows the average path length remains unchanged for the *lookup intensive* workload. This is to be expected, as the lookup intensive workload is dominated by APPLICATION lookups. In the churn intensive workload we see a rise in average hop count as the network size increases; this is because the result was originally dominated by JOIN and MAINTENANCE lookups, which tend to be for closer keys.

1.6.2 Fixing p

In the source of the original model we encountered a minor mistake,[1] which, in many cases, resulted in $p + 1$ parallel requests being generated – rather than the

---

[1]When receiving a timeout or negative response, further queries are dispatched while *pending* $<= p_{max}$, resulting in $p_{max} + 1$ pending queries.

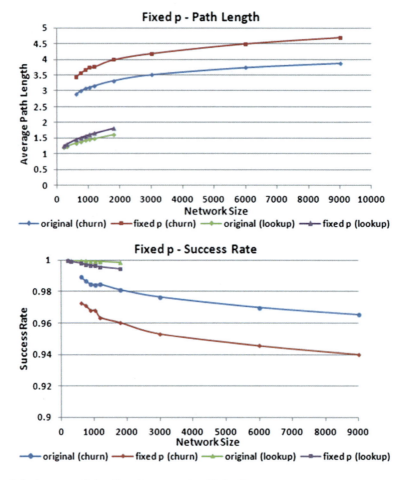

Fig. 1.4 Average path length and success rate with fixed p

supposed maximum of p. Results comparing the average path lengths and success rates when $p = 1$ can be seen in Fig. 1.4.

From these results we observe a rise in average path length, and a small decline in lookup success rate, for both workloads. We also observe a drop in the size of nodes cache tables, which increases with the network size. This is to be expected, as fewer queries are dispatched and hence fewer new nodes are discovered.

1.7 OverSim Results

To match the original scenarios, lookups were performed throughout the entire simulation, with measurements taken from the very beginning. OverSim, by default, only starts performing lookups and recording measurements once the network has reached the desired size, however this is configurable in the settings.

1 An Evaluation of EpiChord in OverSim

Table 1.2 OverSim simulation parameters

Description	Lookup intensive	Churn intensive
Lookup interval	0.5 s	100 s
Network size	{600, ..., 2,000}	{600, ..., 2,000}
Lifetime mean	600 s	600 s
Stabilize delay	60 s	60 s
Cache TTL	120 s	120 s
Cache flush delay	20 s	20 s
Cache check multiplier	3	3
Measurement time	3,000 s	3,000 s
Neighbour list size	4	4
Redundant nodes, l	3	3
Parallelism, p	1, 3, 5	1, 3, 5
Required nodes/slice, j	2	2
Lifetime multiplier, ω	0.5	0.5
Slice multiplier, δ	0.5	0.5

An overview of the simulation parameters can be found in Table 1.2. When we refer to results from the original model, we refer to the results generated after taking the changes in Sect. 1.6 into account. All results are averages of 5 simulation runs.

1.7.1 Finger Cache State

During simulation we measure the average finger cache size for each node, as well as the average accuracy of each node's finger cache. The accuracy is a measure of how many nodes in the finger cache are actually still active within the network.

We observe an average finger cache accuracy of 87 % across all network sizes and both scenarios – almost identical to that of 87.5 % reported in the original paper.

As expected the finger cache size observed in the *lookup intensive* workload is much larger than that in the *churn intensive* workload, due to the extra node information received within lookup messages. The observed finger cache size for varying network sizes under a lookup intensive workload and churn intensive workload can be seen in Fig. 1.5.

1.7.2 Lookup Success Rate

Every lookup performed can be classified into one of four categories:

- Success: The node responsible for the requested key responds positively.
- Failure: No positive response received and no more viable candidates, or reached the maximum hop/time limit
- False-positive: A node has responded positively but is not responsible for the requested key.

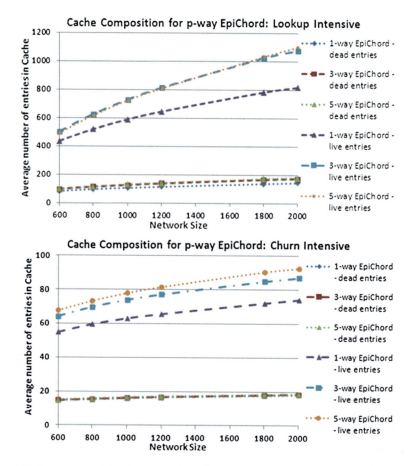

Fig. 1.5 Cache composition for p-way EpiChord under lookup intensive and churn intensive workload

- False-negative: A node did not respond positively but should be responsible for the requested key.

By using false-negative detection, described in Sect. 1.3.3, nodes can detect and handle false-negatives; ultimately they are treated as successful lookups.

The observed success rate for both lookup intensive and churn intensive workloads is shown in Fig. 1.6. Here we use column diagram to show the success rate for *p-way* parallel queries ($p = 1, 3, 5$) for different network sizes up to 2,000 nodes.

As shown in Fig. 1.6, the lookup success rate is marginally higher for lookup intensive workload than for the churn intensive workload. This is expected as under the lookup intensive workload, the larger number of lookups helps to keep the routing state up-to-date whereas for the churn intensive workload, the information propagation rate is lower. Increasing parallelism has only a very slight effect on the success rate. It appears that the lookup improvement is not worth the extra cost of the parallel lookups. The success rates for $p = 5$ is marginally lower than for $p = 3$.

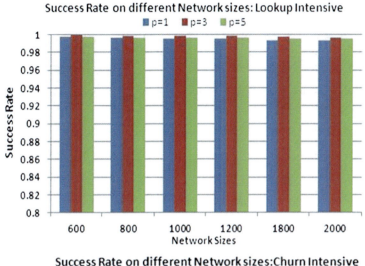

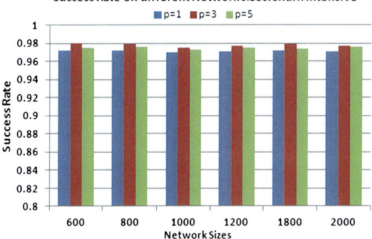

Fig. 1.6 Comparison of success rate for p-way EpiChord for varying network sizes under lookup intensive and churn intensive workload

This rather counter intuitive behavior has also been observed in the original paper and is due to the *5-way* network generating fewer cache-refreshing lookups than a *3-way* EpiChord network.

1.7.3 Lookup Path Length

For each successful lookup performed we also measure the path length – the number of hops taken to find the final destination. Figure 1.7 shows the observed path length for both lookup intensive and churn intensive workloads. We observe

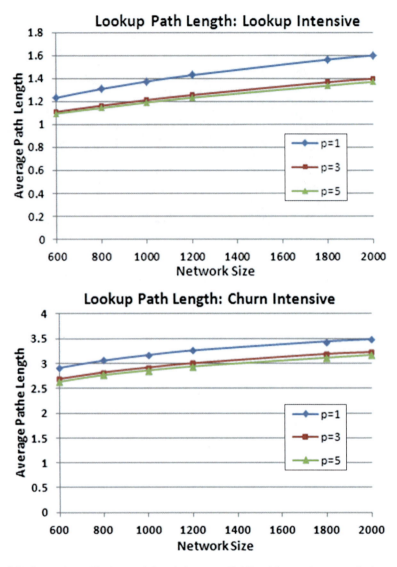

Fig. 1.7 Comparison of lookup path length for p-way EpiChord for varying network sizes under churn intensive and Lookup intensive workload

that in the lookup intensive workload, the hop count varies from 1.1 to 1.4 in both *3-way* and *5-way* EpiChord networks, which signifies that each node has almost complete routing table information and thus allows passing messages nearly in one hop. On the other hand, the hop count varies from 2.8 to 3 under churn intensive workloads with fewer lookups which also satisfies the *O(log n)*-hop lookup performance as depicted in the original paper. Again, the results suggest that an increased level of parallelism in the lookups only marginally improves the hop count, whereas

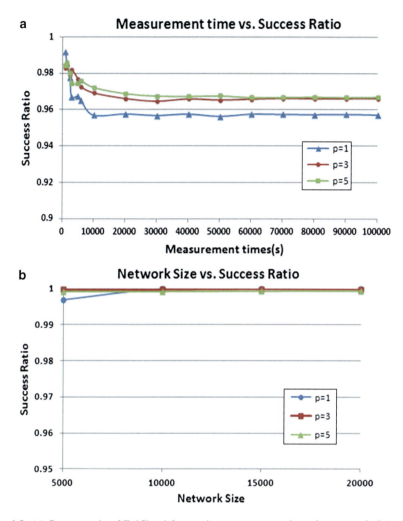

Fig. 1.8 (a) Success ratio of EpiChord for varying measurement times for $p = 1, 3, 5$ demonstrating the stability of the model; (b) Average success ratio of EpiChord for networks with 5,000–20,000 nodes

the increased number of lookups issued in the lookup intensive workload has a much more pronounced positive effect.

1.7.4 Stability and Scalability

We measured the stability of the EpiChord model in OverSim. Figure 1.8a shows the measurement phase vs. success ratio graph for $p = 1, 3, 5$. In OverSim, during the measurement phase, the statistics are collected. Our model has been tested up to 100,000 s and demonstrates that the model is stable after an initial period of between 10,000 s (for p = 1) and 20,000 (for p = 3, 5). EpiChord also has been tested for scalability in terms of network size for scenarios with 5,000–20,000 nodes. Figure 1.8b shows the results for lookup success ratio for different network sizes. This set of results means that the network size does not affect the success rate of EpiChord. As before, p only improves the performance in a rather minor way.

1.8 Conclusion

This paper presented our OverSim EpiChord model, and validated it by comparing our results against the performance of the original EpiChord model. The results for our model closely match those from the original model, supporting the claim that our model is a valid implementation of the EpiChord algorithm. We have then presented amendments to the model and investigated the effects on the performance of the model. Furthermore we have shown that EpiChord and our model in OverSim is stable over an extended period of time. We have also demonstrated that EpiChord achieves excellent results for larger networks. EpiChord's performance is strongly influenced by the number of lookups issued by the nodes as routing table information is attached to lookup return messages. Thus an increased number of lookup message improve the performance of the network, whereas an increased level of parallelism only marginally improves performance. Due to its excellent lookup performance for large scale networks, EpiChord appears well suited to support large distributed environments.

Separately, we have used this model to simulate the effect of different lookup traffic setups, and high node churn to investigate EpiChord's suitability for use in mobile networks [24].

References

1. K. Dhara, Y. Guo, M. Kolberg, X. Wu, Overview of structured peer-to-peer overlay algorithms, in *Handbook of Peer-to-Peer Networking* (Springer, 2009)
2. B. Leong, B. Liskov, E.D. Demaine, EpiChord: parallelizing the chord lookup algorithm with reactive routing state management, in *12th International Conference on Networks 2004 (ICON 2004)*, Singapore, Nov 2004
3. J. Furness, F. Chowdhury, M. Kolberg, EpiChord model for OverSim, http://www.cs.stir.ac.uk/~fch/EpiChord_Model/
4. I. Baumgart, B. Heep, S. Krause, OverSim: a flexible overlay network simulation framework, in *10th IEEE Global Internet Symposium (GI '07)*, May 2007, Anchorage, Alaska, USA
5. The SSFNet project, [Online] Available: http://www.ssfnet.org/. Accessed 01 Aug 2012
6. J. Furness, M. Kolberg, Considering complex search techniques in DHTs under churn, in *2011 I.E. Consumer Communications and Networking Conference (CCNC)*, IEEE, 2011

7. L. Monnerat, C. Amorim, D1HT: A distributed one hop hash table, in *20th IEEE International Parallel & Distributed Processing Symposium (IPDPS)*, 2006
8. A. Gupta, B. Liskov, R. Rodrigues, Efficient routing for peer-to-peer overlays, in *1st Symposium on Networked Systems Design and Implementation (NSDI)*, 2004
9. J. Buford, A. Brown, M. Kolberg, Analysis of an active maintenance algorithm for an O(1)-Hop overlay, in *IEEE Globecom 2007*, Washington DC, USA
10. J. Li, J. Stribling, R. Morris, M.F. Kaashoek, Bandwidth-efficient management of DHT routing tables, in *Symposium on Networked System Design and Implementation (NSDI)*, 2005
11. I. Gupta, K. Birman, P. Linga, A. Demers, R. van Renesse, Kelips: building an efficient and stable P2P DHT through increased memory and background overhead, in *2nd International Workshop on Peer-to-Peer Systems (IPTPS '03)*, 2003, Berkeley, CA, USA
12. I. Stoica, R. Morris, D. Karger, M. Frans Kaashoek, H. Balakrishnan, Chord: a scalable peer-to-peer lookup service for internet applications, in *Conference on Applications, Technologies, Architectures, and Protocols for Computer Communications (SIGCOMM '01)*, ACM, 2001
13. PeerSim P2P Simulator, http://peersim.sourceforge.net. Accessed 05 Jan 2013
14. P2Psim, A simulator for peer-to-peer (P2P) protocols, http://pdos.csail.mit.edu/p2psim/
15. K. Shudo, Y. Tanaka, S. Sekiguchi, Overlay weaver: an overlay construction toolkit. Comput. Commun. **31**(2), 402–412 (2007)
16. PlanetSim: An overlay network simulation framework, http://planet.urv.es/planetsim
17. The Network Simulator – ns-2, http://www.isi.edu/nsnam/ns/
18. A. Brown, M. Kolberg, Tools for peer-to-peer network simulation. Internet-Draft Version 00, IETF, 2006
19. S. Naicken, A. Basu, B. Livingston, S. Rodhetbhai, A survey of peer-to-peer network simulators, in *The 7th Annual Postgraduate Symposium*, Liverpool, 2006
20. D. Stingl, C. Groß, J. Rückert, L. Nobach, S. Kovacevic, R. Steinmetz. PeerfactSim.KOM: a simulation framework for peer-to-peer systems, in *International Conference on High Performance Computing & Simulation (HPCS)*, 2011
21. S. Sioutas, K. Tsichlas, G. Papaloukopoulos, Y. Manolopoulos, E. Sakkopoulos. A novel Distributed P2P Simulator Architecture: D-P2P-Sim, in *ACM International Conference on Information and Knowledge Management (CIKM)*, Hong Kong, 2009
22. F. Dabek, B. Zhao, P. Druschel, J. Kubiatowicz, Towards a common API for structured peer-to-peer overlays. Peer-to-Peer Syst. II **2735**, 33–44 (2003)
23. J. Li, J. Stribling, F. Kaashoek, R. Morris, T. Gil, A performance vs. cost framework for evaluating DHT design tradeoffs under churn, in *INFOCOM*, 2005
24. F. Chowdhury, M. Kolberg, Performance evaluation of EpiChord under high churn, in *The proceedings of the 8th ACM Performance Monitoring, Measurement and Evaluation of Heterogeneous Wireless and Wired Networks (PM2HW2N) Workshop*, Barcelona, 2013

Chapter 2
Enhanced Back-Off Technique for IEEE 802.15.4 WSN Standard

Aditi Vutukuri, Saayan Bhattacharya, Tushar Raj, Sridhar, and Geetha V

Abstract IEEE 802.15.4 is the standard for Low-rate Wireless Personal Area Networks. The CSMA-CA algorithm used in the standard for channel contention causes performance bottlenecks in certain scenarios. We have conducted a performance evaluation of the back-off algorithm with the help of simulations on star networks and identified two parameters which affect the performance of the algorithm – macMinBE and macMaxBE. Further, we have also proposed an enhanced algorithm which involves these two parameters and improves the performance of the back-off algorithm.

2.1 Introduction

2.1.1 Wireless Sensor Networks

A wireless sensor network (WSN) consists of spatially distributed autonomous sensors to monitor and sense physical and environmental conditions [1]. The IEEE 802.15.4 standard aims to provide PHY and MAC layer specifications for ultra low complexity, ultra low cost, ultra low power consumption, and low data rate wireless connectivity among the devices that form a Wireless Personal Area Network.

A. Vutukuri (✉) • S. Bhattacharya • T. Raj • Sridhar • Geetha V
Department of Information Technology, National Institute of Technology,
Surathkal, KA, India
e-mail: aditi.vutukuri@gmail.com; saayan.bhattacharya@gmail.com; tushar.at7@gmail.com; sridhar20061991@gmail.com; geethav.nitk@gmail.com

2.1.2 CSMA-CA Mechanism

CSMA-CA is used by the network nodes to sense the channel and check whether it is idle or not, before transmission of data [2]. Collisions and packet loss are the major problems faced in such a wireless network. This is where the concept of back-off becomes crucial. Back-off algorithm determines how much time should be spent waiting before transmission, when the channel is busy or after collision. A node which backs off for unnecessarily long periods of time can severely hamper the throughput, cause delay and greater energy consumption. Whereas a node which backs off too soon, will perform as many more Clear Channel Assessments (CCAs), find the channel to be still busy and eventually declare failure to transmit. Hence it is important to choose the right method or algorithm for back-off to improve the performance.

The main factors of the back-off algorithm that affect the network efficiency are, the range of the back-off exponent (BE) value, the number of times CCA is performed before transmission, the strategy used to choose and modify the BE value with every successful transmission or collision and the maximum number of times that the node attempts to transmit a packet.

All devices under a personal area network (PAN) coordinator have their back-off periods aligned with it such that every time a device wishes to transmit data it locates the boundary of the next back-off period and waits for a random number of back-off periods. If the channel is busy, following this random back-off, the device waits for another random number of back-off periods before trying to access the channel again. If it finds the channel to be free, then the device locates the next available back-off boundary and transmits the packet. The CSMA-CA mechanism is not used for acknowledgement and beacon frames.

2.1.3 Existing Variations of CSMA-CA Back-Off Mechanism

Several schemes and modifications have been suggested in order to improve the performance of the standard in terms of throughput, delay, energy efficiency or improving other quality of service attributes. Some of these modifications are specifically for the CSMA-CA contention algorithm and focus on reducing time delay between contentions, delays caused due to collisions and improving energy efficiency of the algorithm.

The authors in [3] have proposed two schemes to improve the back-off algorithm so as to better utilize CCAs which are rather energy intensive. Hence, these mechanisms are used to adjust the BE value based on CCA results and packet transmission and in turn also shift the back-off counters to reduce redundant back-offs and CCAs.

In proposed delay mitigation algorithm based on priority [4], the key idea was to reduce the default contention window (CW) value to one for high priority packet. It makes channel access much easier for high priority packets as CCA is performed only once, whereas low priority packets having $CW = 2$ must perform CCA twice. This algorithm greatly reduces delay of a high priority packet.

The state transition back-off scheme in [5] attempts to address the issue of unnecessarily long back-off periods, which reduces the energy efficiency of the mobile stations. When a sensor node first has packet to send, it sets minBE to be 3. After successfully sending each packet, it decreases minBE by 1 until it is 1, if the sensor node still has packet to send. Conversely, it increases minBE by 1, if the sensor node has no data to send in two consecutive superframes, but minBE is limited to be 3.

The dynamic back-off scheme given in [6] proposes a memorized back-off scheme (MBS) along with an exponential weighted moving average (EWMA) approach for dynamic adjustment of the size of the contention window based on the traffic load and the window sizes of the previous successful transmissions. The algorithm outperforms IEEE 802.15.4 during heavy traffic. Though, the authors have solved the problem of collisions due to heavy contention but have not considered the effect of large back-off time between packets.

In the delayed back-off algorithm proposed in [7], different back-offs are allotted to different nodes by the coordinator so as to avoid collision. This is supposed to increase success probability since all the nodes are supposed to finish back-off at different times. However, there is a need to perform 3 CCAs in order to avoid collision between packets using the standard CSMA/CA algorithm and the delayed back-off algorithm. CCA is an energy-intensive activity.

A novel QoS CSMA/CA based on a Gaussian back-off time was proposed in [8] in which the characteristics of Gaussian random variable is changed after every back-off. Further packets of different priorities are supported by maintaining different Gaussian characteristics for each. This scheme is said to be easily adoptable without much change to the standard. However, if the appropriate Gaussian characteristics are not selected, then the collision probability might increase.

We find that the main factors of the back-off algorithm that affect the network efficiency are, the range of the BE value, the number of times CCA is performed before transmission, the strategy used to choose and modify the BE value with every successful transmission or collision and the maximum number of times that the node attempts to transmit a packet. However there is no single modification that is feasible and suitable for all scenarios. While some tackle the problem of collision control, others address the problem of unnecessarily long delays between contentions and yet others require support at the hardware level and are suitable for certain types of applications.

2.2 Enhanced Algorithm

We have conducted a performance evaluation of the back-off algorithm with the help of simulations on star networks and identified two parameters which affect the performance of the algorithm – macMinBE and macMaxBE. Further, we have also proposed an enhanced algorithm which involves these two parameters and improves the performance of the back-off algorithm.

2.2.1 Motivation

Wireless sensor networks are used in several critical time-bound applications, where data from every node (mote) is crucial and the failure to receive data from all nodes can cause severe performance loss or in certain cases potential damage. Some of these applications include forest fire detection, sensing leaks in nuclear reactors, biomedical sensing and other such fault intolerant applications. Our focus has been on primarily such high data rate applications and a mechanism to minimize loss of packets due to excessive contention and collision.

According to the IEEE 802.15.4 standard, the BE value is set to a value in the range of two variables macMinBE and macMaxBE. The macMaxBE (henceforth referred to as simply maxBE) value indicates the maximum value of the back-off exponent BE. The macMinBE (henceforth referred to as simply minBE) value is the minimum value of the back-off exponent. In order to enhance the performance of the back-off algorithm, we suggest that these values be flexible and dynamic to changes. Further, these values must be changed by keeping in mind the maximum number of back-offs allowed for a particular packet. This value is decided by the maxCSMABackoffs parameter. By default, this value is set to 5. Hence, the parameters must be changed such that the range of values for BE has exactly 5 values. It is important that a different BE value be chosen in all the attempts by a node to send a packet so that the probability of choosing a different wait time each time increases.

Changes to these parameters must be performed according to some agenda on how to handle failures and successes in transmissions. Our agenda involves giving priority to the failures in transmission. Nodes which face the problem of consecutive failures in transmission, will choose a BE value from a range such that it needs to spend lesser time backing off (waiting) and instead trying more often to get the channel.

2.2.2 Suggested Changes

The following changes were proposed based on our analysis.

- Initially minBE is set at 2 and maxBE is set at 6 for every node.
- Following two consecutive failed transmissions for each node, minBE and maxBE are both decremented by 1. This is so that any node with two consecutive failed transmissions is allowed to contest for the channel more vigorously.
- Following two consecutive successful transmissions, the minBE and maxBE values are incremented so that it contests for the channel less vigorously, thereby allowing other nodes to successfully transmit their packets.
- The minimum limit for minBE is set at 1 and maximum value at 3(default value in standard CSMA-CA). The minBE value is the first value BE assumes in order to wait while attempting to transmit a packet. A wait is essential to keep the collision probability under check. A minimum value of 1 is chosen so that in

2 Enhanced Back-Off Technique for IEEE 802.15.4 WSN Standard

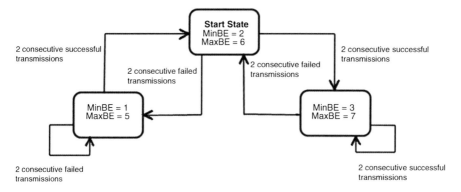

Fig. 2.1 The proposed state transition based back-off algorithm

case of repeated failed transmissions, a node performs a short wait before sending a packet. Similarly, an upper limit of minBE at 3 is put in place so that if a node has been successful in its past few transmissions, its minBE does not increase to a number such that it has to wait unnecessarily long periods to transmit a packet. The limits on the maxBE value have been put accordingly so that a node can have its BE value assume a different value each time it finds the channel to be busy. Hence the minimum limit for maxBE value is set at 5 and maximum limit at 7.

Figure 2.1 depicts the changes we have proposed in this algorithm.

2.2.3 Simulation Results

NS2 (Network Simulator 2) is a network simulation software that allows us to design Wireless Sensor Networks and analyze the performance of the network. We used the NS2 simulator to perform our experiments and modified the simulator's implementation of the CSMA-CA algorithm in order to observe the performance of our enhanced algorithm (Table 2.1).

Our main intention with this scheme was to address the issue of the number of packets dropped due to successive collisions. The standard uses the maxCSMABackoffs parameter to determine the maximum number of retries allowed for a packet. The following graph shows the number of packets that are dropped on reaching maxCSMABackoffs.

Figure 2.2 shows how the number of packets dropped on reaching maxCSMABackoffs varies for both the standard algorithm and the enhanced algorithm as the number of nodes are increased. As seen, the number of packets dropped is far lesser, by about 55.6 %, for our enhanced algorithm when compared to the standard and this trend is observed for all the variations in the number of nodes.

Table 2.1 Simulation parameters – star topology

Traffic type	CBR traffic
Packet size	70 bytes
Data interval	0.6 s
Number of nodes	3–15
Routing type	AODV
Topology	Star topology
Antenna type	Omni-directional antenna

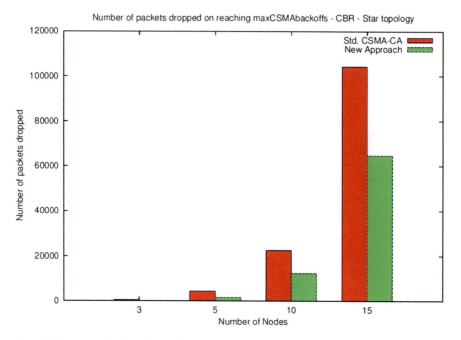

Fig. 2.2 Number of packets dropped versus number of nodes

The other network parameters used to test the performance of the enhanced algorithm against the standard are end-to-end delay, throughput and the number of packets sent.

Figure 2.3 shows the variation of end-to-end delay versus number of nodes. End-to-end delay is calculated as the average amount of time the packets take to reach from the sensor nodes to the PAN coordinator. As the number of nodes increases the difference between the end-to-end delays of the standard algorithm and the enhanced algorithm becomes more prominent. The enhanced algorithm tends to exhibit lesser end-to-end delay compared to the standard algorithm in all cases. The average decrease in the End-to-End delay is about 12 % when compared to standard back-off algorithm.

Figure 2.4 shows how the number of sent packets vary for both the algorithms as the number of nodes is increased. The dynamism and the flexibility in the enhanced

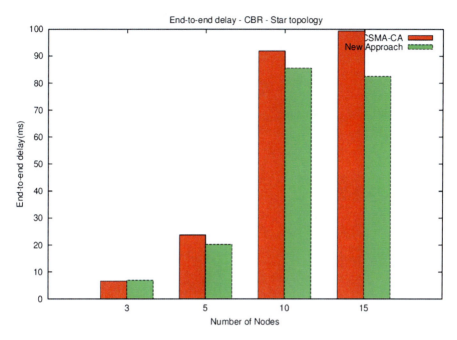

Fig. 2.3 End-to-end delay (ms) versus number of nodes

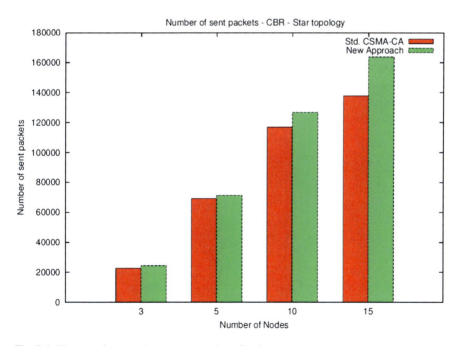

Fig. 2.4 Number of sent packets versus number of nodes

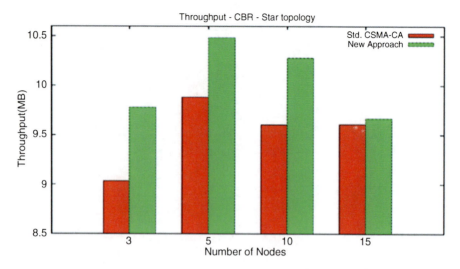

Fig. 2.5 Throughput versus number of nodes

scheme are helpful in sending a greater number of packets in a limited time period. We have observed an increase of 9.47 % on an average in the number of packets sent using the enhanced algorithm.

Throughput is a direct measure of how many packets are received at the PAN coordinator. Figure 2.5 indicates the performance of the algorithms in terms of throughput. The throughput of the network is more when using the enhanced scheme amounting to an increase 5.24 % when compared to the standard CSMA-CA back-off algorithm, however, the receiving capability of the PAN coordinator poses a limitation on the amount of packets received as the number of nodes is increased.

2.3 Conclusion

The enhanced scheme effectively adjusts the range from which the BE value is chosen keeping in mind the failures and successes in packet transmissions. The proposed algorithm was, however, found to be more suitable for all high density traffic in a wireless sensor network. In high-data rate scenarios, where nodes are sending packets often, failure in transmission of packets is a common problem. It reduces the end-to-end delay which helps in critical real time systems with time bound performance requirements, for example, in a forest fire detection scenario, or wild life monitoring scenarios. The throughput is also fairly high compared to the CSMA-CA algorithm which signifies that the rate at which the sink receives the data is also improving. We have observed a significant increase in the number of sent packets. The only limiting factor is the receiving capability of the sink. Moreover we also observe a decrease in the number of packets dropped because of exceeding the NB or number of back-offs. This shows that the channel contention itself is performing better in these scenarios than in CSMA-CA algorithms. Future work in this area involves analyzing the energy efficiency of this algorithm against the standard CSMA-CA algorithm.

References

1. IEEE 802.15.4, Part 15.4, *Low-rate wireless personal area networks (LR-WPANs)*. (IEEE Standard for Local and Metropolitan Area Networks, 2006)
2. S. Tariq, *MAC algorithms in wireless networks: applications, issues and comparisons*. Master thesis, Department of Computing Science, Umea University, Sweden, 2005
3. J.Y. Ha, T.H. Kim, H.S. Park, S. Choi, W.H. Kwon, An enhanced CSMA_CA algorithm for IEEE 802.15.4 LR-WPANs. IEEE Commun. Lett. **11**(5), 461–463 (2007)
4. T.H. Kim, S. Choi, Priority-based delay mitigation for event monitoring IEEE 802.15.4 LR-WPANs. IEEE Commun. Lett. **10**(3), 213–216 (2006)
5. J. Ko, Y.H. Cho, H. Kim, Performance evaluation of IEEE 802.15.4 MAC with different back-off ranges in wireless sensor networks, in *IEEE International Conference on Communication Systems,* Singapore, 2006
6. A.C. Pang, H.W. Tseng, Dynamic back-off for wireless personal networks. IEEE GLOBECOM 2004 **3**, 1580–1584 (2004)
7. B.H. Lee, H.K. Wu, A delayed back-off algorithm for IEEE 802.15.4 beacon-enabled LR-WPAN, in *Proceedings of 6th International Conference on Information, Communications and Signal Processing*, 2007, pp. 1–4
8. M.J. Youn, Y.Y. Oh, J. Lee, Y. Kim, IEEE 802.15.4 based QoS support slotted CSMA/CA MAC protocol for wireless sensor networks, in *International Conference on Sensor Technologies and Applications*, 2007, pp. 113–117

Chapter 3
Efficient Retransmission QoS-Aware MAC Scheme in Wireless Sensor Networks

M. Kumaraswamy, K. Shaila, V. Tejaswi, K.R. Venugopal, S.S. Iyengar, and L.M. Patnaik

Abstract In this paper, an Efficient Retransmission Random Access Protocol (ERRAP) is designed that combines scheme of collision avoidance and energy management for low-cost, short-range wireless radios and low-energy sensor nodes applications. This protocol focuses on efficient Media Access Control (MAC) schemes to provide autonomous Quality of Service (QoS) to the sensor nodes in one-hop QoS retransmission group in WSNs where the source nodes do not have receiver circuits. These sensor nodes can only transmit data to a destination node, but cannot receive acknowledgement or control signals from the destination node. The proposed scheme ERRAP provides QoS to the nodes which work independently on predefined time by allowing them to transmit each packet an optimal number of times within a given period. Our simulation results demonstrate the superiority of ERRAP scheme which increases the delivery probability and reduces the energy consumption.

3.1 Introduction

Wireless Sensor Networks (WSNs) consist of a large number of distributed nodes, that combine automated sensing, embedded computing and wireless components into tiny embedded devices. WSNs gather information or detect special events and communicate in a wireless fashion. WSNs provide a wide range of potential

---

M. Kumaraswamy (✉) · K. Shaila · V. Tejaswi · K.R. Venugopal
Department of Computer Science and Engineering, University Visvesvaraya College of Engineering, Bangalore University, Bangalore, India
e-mail: kumarasm67@yahoo.co.in

S.S. Iyengar
Florida International University, Miami, FL, USA

L.M. Patnaik
Indian Institute of Science, Bangalore, 560 001, India

applications like healthcare, environmental monitoring, battlefield monitoring, remote sensing, industrial process control, surveillance and security etc. A typical Wireless Sensor Network consists of one or few sink nodes and a large number of data sensor nodes deployed, each source node generates data and transmits to the sink through common communication channel. In general, such networks consists of both transmitting and receiving circuits. However, in most applications the devices generally collect data and transmit to the sink node. The communication from sink node to source node is minimum. The receiver circuitry adds extra cost and also consumes significant amount of energy during process. Thus by using sensor data nodes with only transmitters, the device cost and the entire network infrastructure cost can be reduced. These wireless sensor devices are equipped with sensing, computation and wireless communication capabilities. Sensing tasks for sensors devices could be temperature, light, sound, humidity, vibration, etc. Random access scheme WSNs communication devices are low rate communication protocol are designed for low cost, low data rate and low power WSNs devices.

Optimal retransmission is the process of sending packets to the sink multiple number of times to achieve the maximum delivery probability. The efficient retransmission in WSNs is mainly focused on QoS in terms of packet delivery probability and energy efficiency.

In this paper, we consider WSNs that deals with QoS aware medium access control scheme of one-hop QoS group that has low complexity, less power consumption and optimum cost. Proposed scheme consists of network topology with more number of source nodes which are distributed and decentralized in one-hop communication range. In the present day environment every source node in a WSN is equipped with only transmitter module. The receiver module is avoided, since they consume more energy and are expensive due to the hardware complexity. The throughput requirement are low because the source collects and transmits the data to the sink. The sink node in the network has both transmitter and receiver and it receives the data transmitted by the source nodes. There are a large number of applications which use the above concept such as Smart Home Monitoring, Smart Environment, Intelligent Transportation and Medical Monitoring.

Most of the medium access control protocol are like polling, CSMA, Automatic Repeat Request (ARQ), collision avoidance/detection [1] and scheduled transmissions [2] are not effective because they need the ACK to transmit the next packet.

Motivation In many application scenarios of sensor networks, sensor data must be delivered to the sink node within time constraints. It is crucial to evaluate the performance limits, such as maximum data delivery and energy consumption of traffic loads under all conditions.

Hence, efficient retransmission, maximization of the packet delivery probability and energy efficiency have to be considered in designing WSNs.

Contribution This paper presents, Efficient Retransmission Random Access Protocol (*ERRAP*) provides QoS to the nodes using random access mode where each node transmits data packet by selecting variable slot randomly for adaptive data packet considering local environment. Nodes can only join the network during random access periods. The time interval between random access periods could be small. So, in the proposed protocol, nodes randomly decide whether it should retransmit to improve packet delivery depending on some pre-calculated efficient retransmission probabilities. The sink node receives exactly one error-free retransmission data packet without collision.

An analytical method to evaluate the maximum data delivery probability and minimum energy consumption is proposed. We design an Efficient Retransmission Random Access Protocol (*ERRAP*), is simple, lightweight, compatible with the 802.11b standard and provides maximum data delivery.

Organization The rest of the paper is structured as follows. Related works are discussed briefly in Sect. 3.2. An overview of the Background is given in Sect. 3.3. In Sect. 3.4 describes the Problem Definition, objectives and assumptions. Section 3.5 presents the System Model. Mathematical Model is developed for the One-Hop Retransmission in Sect. 3.6. Algorithm and Performance Evaluation are presents in Sects. 3.7 and 3.8. Conclusions are contained in Sect. 3.9.

3.2 Related Work

Pai et al. [3] designed a novel adaptive retransmission algorithm to improve the misclassification probability of distributed detection with error-correcting codes in fault-tolerant classification system for Wireless Sensor Networks. The local decision of each sensor is based on its detection result. The detection result must be transmitted to a fusion center to make a final decision.

Lu et al. [4] proposed a MAC layer cooperative retransmission mechanism and a node can retransmit lost packets on behalf of its neighboring node. However, although each lost packet can be recovered by a neighboring node, it still requires a new transmission for each retransmission attempt, which largely limits its ability to increase the throughput of the network.

Xiong et al. [5] consider cooperative forwarding in WSNs from a MAC-layer perspective, which means a receiver can only decode one transmission at a time. Qureshi et al. [6] propose a latency and bandwidth efficient coding algorithm based on the principles of network coding for retransmitting lost packets in a single-hop wireless multicast network and demonstrate its effectiveness over previously proposed network coding based retransmission algorithms.

Ruiz et al. [7] propose an architecture collaboration in which the MAC and routing protocols discover and reserve routes to organize nodes into clusters and to schedule the access to the transmission medium in a coordinated time-shared fashion. It achieves QoS and reduces energy consumption by avoiding collisions and considerably lowering idle listening.

Tannious et al. [8] have proposed an algorithm where a secondary node user exploits the retransmissions of primary node user packets in order to achieve a higher transmission rate. The secondary node receiver can potentially decode the primary node users packet in the first transmission.

Bai et al. [9] propose a design of IEEE 802.11 based wireless network for MAC that dynamically adjusts the retransmission limit to track the optimal trade-off between transmission delay and packet losses to optimize the overall network control system performance.

He and Li [10] propose the single-relay Cooperative Automatic Repeat Request (CARQ) protocol, in which the relay node is selected in a distributed manner; the relays use different backoff time before packet retransmission. In a dense network, due to high possible collision probability among different contending relays, an optimized relay selection scheme is introduced to maximize system energy efficiency by reducing collision probability.

Volkhausen et al. [11] focuses on cooperative relaying, that exploits temporal and spatial diversity by additionally transmitting status a relay node; such relaying improves packet error rates and transmit only once rather than on each individual hop along the routing path. This cooperation reduces the total number of transmissions and improves overall performance.

Wang et al. [12] propose the local cooperative relay for opportunistic data forwarding in mobile ad-hoc networks. The local cooperative relay select the best local relay node without additional overhead; such real time selection can effectively bridge the broken links in mobile networks and maintain connectivity.

3.3 Background

Sudhaakar et al. [13] propose a Medium Access Control scheme, which typically consists of large number of source nodes and are within one-hop communication range to one or few sink nodes. Each of these source nodes is equipped with only transmitter module in order to eliminate the cost due to hardware complexity and energy consumption of the receiver module. As a result, they are not capable of receiving any signals like ACK/NAK. The source nodes collect data and transmit a relatively small data frame to the sink nodes once a while and hence the throughput requirement of the source nodes is low. The sink nodes are the only nodes in the network that are equipped with receiver modules and are capable of receiving the transmissions of the source nodes.

3.4 Problem Definition

Consider a Wireless Sensor Network consisting of M nodes as shown in Fig. 3.1, having source nodes and one or two sink nodes. All the WSNs nodes are within one hop transmission range of the sink. The source nodes do not have receiver unit, so it is impossible for sensor nodes to sense the channel either for collision detection or receive any acknowledgements from the sink node. The main objective of the proposed work is to

Fig. 3.1 Nodes deployment in wireless sensor network

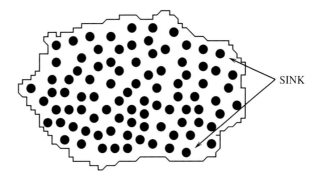

1. Maximize packet delivery probability.
2. Reduce packet retransmission.
3. Reduce energy consumption.

3.4.1 Assumptions

(i) All nodes are homogeneous.
(ii) The nodes are randomly distributed within a area.
(iii) Packet generation rate at each node follows a Poisson distribution.

3.5 System Model

The *ERRAP* provides QoS to the nodes using random access mode where each node independently depending on local conditions transmits data packet by selecting variable slot randomly for variable data packet according to the number of sensors nodes. Nodes can only join the network during random access periods. The time interval between random access periods could be small. In the proposed protocol *ERRAP*, nodes randomly decide whether it should retransmit to help the packet delivery depending on some pre-defined efficient retransmission probabilities. The sink node receive exactly one error-free transmission packet in a slot, without collision. Our goal is to develop decentralized MAC protocol to provide QoS guarantees for both time-critical and non time-critical sensor applications. This is a challenge that has not been addressed by any existing approach. The most important metrics to analyze the QoS performance of MAC protocol is packet delivery probability and energy efficiency.

In WSNs, each sensor node data packets transmission duration is relatively small when compared to the data packets that are generated at a constant rate i.e., one packet every *T* units of time. In addition, if a packet cannot be successfully delivered within a data generation *T* units of time, the data packets is simply neglected. This makes sures that the new data packets have greater chance of being successfully delivered. Thus the maximum delivery probability that can be achieved by each individual sensor node increases eventually, so that all the nodes in the network achieve their required QoS in terms of data delivery probability.

3.6 Mathematical Model

3.6.1 One-Hop Retransmission

We have assumed that the source nodes generate data at constant rate of one packet every *T* units of time and the retransmission time for each packet is much smaller than the duration of packet transmission T_p. To achieve equal packet delivery probability by all the nodes in the WSNs. Under this assumption the packet arrival rate can be modeled as a Poisson distribution. The number of nodes in the network is denoted by *M* and the number of retransmissions by each node for each packet is denoted by y_k. The notations are defined in Table 3.1.

The packet arrival rate of the source nodes can be modeled as a Poisson distribution and the probability that *p* packets are transmitted in an interval T_t with *Q(M)* the probability of *M* arrivals in one time slot is given by

$$Q(M) = \frac{(\beta T_t)^p}{p!} e^{-\beta T_t} \quad (3.1)$$

Where, β is the rate of traffic generated by all other nodes inside the transmission range of a node and is equal to $\frac{(M-1)}{T} y$.

The probability that the packet transmitted by node *k* does not collide, so it is same as the probability that no packet were transmitted by the other $M - 1$ nodes in an interval $2T_p$. Therefore Q_{nc} is

$$Q_{nc} = e^{-2\beta T_p} \quad (3.2)$$

Table 3.1 Notations

Symbols	Meaning
M	Total number of sensor nodes
T	Data packet generation time
T_p	Duration of packet transmission
y_k	Number of retransmission
β	Packet arrival rate
p	Number of packets
Q	Packet delivery probability
T_t	Time of *p* packet transmission

3 Efficient Retransmission QoS-Aware MAC Scheme in Wireless Sensor Networks

The above discussion presents the probability that a packet transmitted by node k is successfully received by the sink. However, node k transmits y_k copies of the packet at random instants in every time interval T_p. Hence the actual parameter of interest will be the probability that at least one of these y copies is successfully received at the sink, which is defined as the QoS delivery probability of the node. The $Q(y_c)$, the collision transmission of delivery probability of each packet is given by

$$Q(y_c) = (1 - Q_{nc})^y \tag{3.3}$$

The probability of successful transmission of sensor data packet $Q(y_s)$ is given by

$$Q(y_s) = (1 - Q(y_c)) \tag{3.4}$$

Combining the above equations, we obtain

$$Q(y_s) = (1 - Q(y_c))$$

$$Q(y_s) = 1 - (1 - Q_{nc})^y$$

$$Q(y) = Q(y_s) = 1 - (1 - e^{-2\beta T_p})^y \tag{3.5}$$

The $Q(y)$ expresses the QoS delivery probability as function of the number of retransmissions attempted by each node in the interval T_p.

The maximum delivery probability Q_{max} that can be achieved is given by

$$Q_{max} = 1 - (1 - e^{-2\beta T_p})^y \tag{3.6}$$

The above result gives relationship between the maximum delivery probability that can be achieved, the number of retransmission attempts that each node makes in every interval T_p and the number of nodes M.

3.7 Algorithm

In this paper, the performance of retransmission algorithm *ERRAP* is discussed to find the solution to the optimization problem in one-hop QoS group containing M nodes. The objective of the *ERRAP* algorithm is to find the optimal retransmission value between y_{low} and y_{high} that minimizes the total sensor network traffic and each node in WSNs achieves maximum delivery probability in background traffic.

ERRAP algorithm solves the problem of energy consumption and delivery probability.

The sensor nodes are randomly deployed and sends data packet to a sink node, y number of times. All the WSNs nodes are within one hop transmission range of the sink and source nodes in the network to achieve the same delivery probability as shown in Algorithm 1. When a data packet is received by a node after transmitting y number of times, in a given period. The arrival rate of packet is modeled as a poisson distribution and the maximum delivery probability is achieved in one hop retransmission.

Algorithm 1: EERAP algorithm

Begin

1. All M Sensor Nodes are within the Sensing and Communication Range
2. Nodes are Randomly Distributed
3. All Source Nodes Send Data Packet to a Sink Node
4. Sink has information about each Source Node Location and ID
5. Each Node Energy Depends on Distance and Data Size
6. Each Source Nodes are Transmits y copies at Random Instant
7. Data Packets are received within the given time Period at Sink
8. Delay from each Source Node to a Sink is same
9. Probability of Error is Minimized
10. Minimum Number of Retransmission y times from one-hop QoS Group
11. Maximum Delivery Probability $Q_{max} \leftarrow 1 - (1 - e^{-2\beta T_p})^y$

End

3.8 Performance Evaluation

3.8.1 Simulation Setup

The performance of *ERRAP* has been evaluated using ns2 simulator package to obtain packet delivery probability and energy consumption. A random flat-grid scenario is chosen for deployment of the nodes within 50×50 and 230×230 m area. In our simulation model, we use two-ray ground reflection model for radio propagation and omnidirectional antenna. The transmission bandwidth is set to 50 Mbps, each source node has only transmit circuit and no receiver and the number of nodes M is 100, data arrival rate $T = 1$ ms and packet transmission time $T_p = 6.4 \times 10^{-4}$ ms.

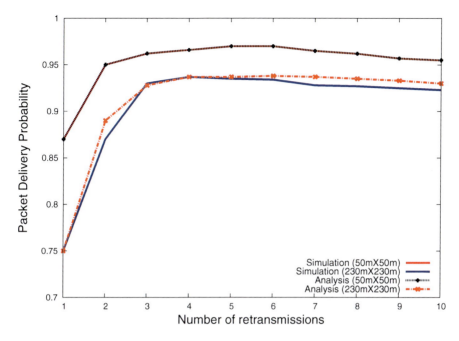

Fig. 3.2 Packet delivery probability of one-hop retransmission simulation and analysis result

3.8.2 One-Hop QoS Group

The results for $Q(y)$ given in Eq. 3.4 for one-hop retransmission, consisting of $M = 100$ nodes is plotted in Fig. 3.2. It shows that the probability of the delivery of packets initially increases with the number of retransmissions, reaches maximum and then decreases. The simulation and numerical analysis results shows that the maximum delivery probability of $Q(y)$ is 0.9990 for 50×50 m area when $y = 4$ or $y = 5$. The minimum delivery probability $Q(y)$ is 0.978 is achieved for $3 \leq y \leq 9$. The *ERRAP* scheme minimizes the network traffic when $y = 3$ and maximizes the probability of delivery of data packets when the retransmission value $y = 4$.

The second set of curves of *ERRAP* of simulation results is comparable with the theoretical analysis. The delivery probability $Q(y)$ is 0.96 when the nodes are randomly distributed in 230×230 m region. Since, the simulation performance of sensor nodes are poor in a large region, we assume that the packet loss is only due to channel errors and not due to collisions or interference.

The graph in Fig. 3.2 illustrates that the number of retransmission by each sensor node is reduced by choosing the value for y as 3 or 4. This increases the probability of delivery, which in turn increases the lifetime of the sensor nodes. The simulation results are agree with the analytical results.

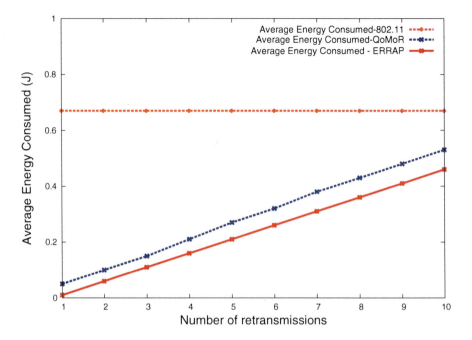

Fig. 3.3 Average energy consumption of one-hop retransmission and comparison

3.8.3 Minimizing Energy Consumption

The goal of *ERRAP* scheme is to minimize the energy consumption between any source node and the sink node. The lifetime of sensor nodes in the network is directly proportional to the energy dissipation of each sensor node. The consumed energy in sensors includes the energy required for sensing, transmitting, receiving and processing of data. *ERRAP MAC* scheme contribute to energy efficiency by minimizing collisions and retransmissions. We simulate the performance of *ERRAP* scheme with respect to energy consumption and compare the average energy consumption with *QoMoR* and *802.11b*.

The goal of the collision avoidance scheme is based on medium access control of the *ERRAP* to increase the channel access probability for fairly distributing the energy consumption of the stations and thereby increasing the network lifetime. Figure 3.3 shows the average energy consumption of the *ERRAP* scheme for different values of retransmissions with 2 K data packets when the aggregated data rate generated by all the nodes is about 50 Mbps, which is equal to the available bandwidth. The energy consumed by the *ERRAP* scheme for the number of retransmissions value 10, is less than the energy consumed by the *QoMoR* and *802.11b* protocol. The *ERRAP* scheme uses shorter frame slots, avoiding control packets like *RTS* and *CTS*, which unnecessarily consume energy and bandwidth.

3 Efficient Retransmission QoS-Aware MAC Scheme in Wireless Sensor Networks

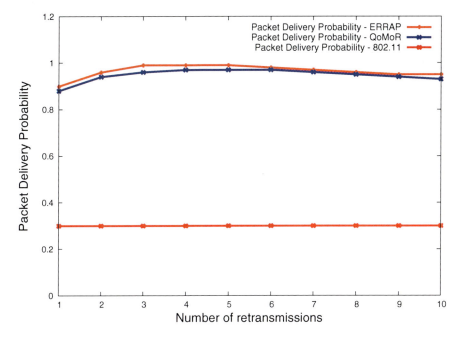

Fig. 3.4 Packet delivery probability of one-hop retransmission and comparison

Figure 3.4 shows the packet delivery probabilities achieved by *ERRAP* scheme, *QoMoR* and *802.11b* under the same conditions. The *ERRAP* protocol is significantly higher than that achieved by the *QoMoR* and *802.11b* protocol. Both *QoMoR* and *802.11b* do not use the available bandwidth as efficiently as *ERRAP*. The *ERRAP* provides QoS to the nodes using random access mode where each node transmits data packet by selecting variable slots randomly for variable data packet according to the total number of sensors nodes and each node take local decisions, depending on some pre-defined efficient retransmission probabilities.

When the number of nodes is large and the aggregate data rate is matching the available channel bandwidth, the performance of the *ERRAP* scheme is significantly better than *QoMoR* and *802.11b* both in terms of *QoS*, delivery probability and energy consumption for the event-driven applications.

3.9 Conclusions

In this paper, we have designed and proposed an Efficient Retransmission Random Access Protocol (*ERRAP*) algorithm scheme which is a combination of collision avoidance and energy management for short-range, low-cost and low-energy WSN's applications like Smart Environment, Home Automation, Structure Monitoring, Intelligent Transportation and Medical Monitoring. In our work, we have assumed that the source nodes do not have receiver circuits. Hence, they can only transmit data packet to destination node, but cannot receive any acknowledgement control signals from destination. In *ERRAP* scheme, each source node simply retransmits each of its data packet an optimal number of times within a given period of time in one-hop QoS group. The source nodes employ probabilistic retransmission to minimize the energy consumption and maximize the packet delivery probability. In future work, we will focus on Two-QoS groups for both analytical and simulation to compare our scheme with other schemes.

References

1. K. Jamieson, H. Balakrishnan, Y.C. Tay, *Sift: A MAC Protocol for Event-Driven Wireless Sensor Networks* (MIT Laboratory Computer Science, Cambridge, 2002)
2. W.Ye.J. Heidenmann, D. Estrin, Medium access control with coordinated adaptive sleeping for wireless sensor networks. IEEE/ACM Trans. Netw. **12**(3), 493–506 (2004)
3. H. Pai, J. Sung, Y.S. Han, Adaptive retransmission for distributed detection in wireless sensor networks, in *Proceedings of IEEE International Conference on Sensor Networks, Ubiquitious, and Trustworthy Computing*, vol. 2, Taichung, 2006, pp. 2–7
4. K. Lu, S. Fu, Y. Qian, Increasing the throughput of wireless LANs via cooperative retransmission, in *Proceedings of Globecom'07*, Washington, DC, 2007, pp. 5231–5235
5. L. Xiong, L. Libman, G. Mao, Optimal strategies for cooperative Mac-layer retransmission in wireless networks, in *Proceedings of IEEE Wireless Communications and Networking Conference*, Las Vegas, NV, 2008, pp. 1495–1500
6. J. Qureshi, C.H. Foh, J. Cai, An efficient network coding based retransmission algorithm for wireless multicast, in *Proceedings of IEEE Personal Indoor and Mobile Radio Communications*, Tokyo, 2009, pp. 691–695
7. J. Ruiz, J. Gallardo, L. Villasenor Gonzalez, D. Makrakis, H. Mouftah, QUATTRO: QoS-capable cross-layer MAC protocol for wireless sensor networks, in *Proceedings of IEEE Global Telecommunications Conference*, Honolulu, 2009, pp. 1–6
8. R.A. Tannious, A. Nosratinia, Cognitive radio protocols based on exploiting hybrid ARQ retransmissions. IEEE Trans. Wirel. Commun. **9**(9), 2833–2841 (2010)
9. J. Bai, E.P. Eyisi, Y. Xue, X.D. Koutsoukos, Dynamic tuning retransmission limit of IEEE 802.11 MAC protocol for networked control systems, in *Proceedings of 3rd IEEE/ACM International Conference on Cyber, Physical and Social Computing (CPSCom)*, Hangzhou, 2010, pp. 666–671
10. X. He, F.Y. Li, An optimal energy efficient cooperative retransmission MAC scheme in wireless networks, in *Proceedings of IEEE International Conference on Wireless Vehicular Technology, Information Theory, Aerospace and Electronic System Technology*, Chennai, 2011, pp. 1–5
11. T. Volkhausen, K. Dridger, H.S. Lichte, H. Karl, Efficient cooperative relaying in wireless multi-hop networks with commodity WiFi hardware, in *Proceedings of 10th IEEE International Symposium on Modeling and Optimization in Mobile, Ad Hoc and Wireless Networks (WiOpt)*, Paderborn, 2012, pp. 299–304
12. Z. Wang, C. Li, Y. Chen, Local cooperative relay for opportunistic data forwarding in mobile Ad-hoc networks, in *Proceedings of IEEE International Conference on Communications (ICC)*, Ottawa, 2012, pp. 5381–5386
13. R.S. Sudhaakar, S. Yoon, J. Zhao, C. Qiao, A novel QoS-aware MAC scheme using optimal retransmission for wireless networks. IEEE Trans. Wirel. Commun. **8**(5), 2230–2235 (2009)

Chapter 4
Performance Analysis of Correlated Channel for UWB-MIMO System

Mihir Narayan Mohanty, Monalisa Bhol, and Sanjat Kumar Mishra

Abstract Ultrawideband (UWB) communication has attracted great interest during the last two decades. Simultaneously, MIMO systems occupied the same attraction. Due to high data rate communication the spectral efficiency increases in spatially multiplexed (SM) multiple-input multiple-output (MIMO) systems. In this paper, the concept and techniques of MIMO has been extended to UWB systems. But, the impact of the spatial correlation requires estimation on the performance of the spatial multiplexing SM-UWB-MIMO system. Another issue is the performance degrades in the presence of high values of spatial correlation. To mitigate such problem, researchers have worked with various methods. Though it has been done to some extent, still it has been designed for virtual UWB-MIMO Time Reversal (TR) system for verification in initial stage. Another novel method also introduced to reduce the effect of correlation. It has been chosen by taking the Eigen value of the channel matrix for the computation of the system performance. The result shows its performance.

4.1 Introduction

Ultrawideband (UWB) communication system has become most promising for high data rate as well as short-range communication systems. Therefore, it has attracted great interests from both academic and industrial aspects recently. UWB communications are suitable for short-range communications due to the consumption of less transmitting power. Some of the examples of this application are in the field of sensor networks and personal area networks (PANs) [1].

M.N. Mohanty (✉) • M. Bhol
ITER, Siksha 'O' Anusandhan University, Bhubaneswar, Odisha 751030, India
e-mail: mihir.n.mohanty@gmail.com; monathegreat33@gmail.com

S.K. Mishra
Seemanta Engineering College, Jharpokharia, Mayurbhanj, Odisha, India
e-mail: sanjatmishra@seemantaengg.ac.in

Increasing demand for higher wireless system capacity has catalyzed several transmission techniques, among which multiple-input/multiple-output (MIMO) technology is popular one. Extending MIMO technology to the UWB regime, a large gain in the channel capacity, robustness and coverage radius is noticed in UWB indoor communications systems [2]. MIMO systems are well-equipped with multiple numbers of antennas, at both the transmitter and receiver or the array antenna in order to improve performance.

MIMO technologies overcome the deficiencies of the traditional methods through the use of spatial diversity. Data can be transmitted over M transmit antennas to N receive antennas supported by the receiver terminal. Such systems are used in wireless communication for enhancement of capacity and bit error rate (BER). It offers significant increases in data throughput and link range without additional bandwidth or transmit power. These characteristics are necessary for the future generation of Telecommunications systems. Rayleigh fading has been considered as the propagation channel for verification. Diversity gain and spatial multiplexing (SM) are the two main advantages of MIMO systems that are used to study the effect of increase in bit rate with increasing the number of transmitter and receiver antennas. In MIMO system, we primarily need to take into account the spatial correlation. The effect of spatial correlation has to be minimized to obtain better system performance. In [3], the time-reversed channel impulse response (CIR) is implemented as a filter at the transmitter side. It is well known that the MIMO-TR-UWB system can achieve transmit diversity, but it suffers from both transmit and receive antenna correlations. The single-input multiple-output TR-UWB (SIMO-TR-UWB) or virtual MIMO-TR-UWB does not face the transmit antenna correlation because it has only one transmit antenna.

4.2 Related Literature

An overview of reported measurements and modelling of the UWB indoor wireless channel is presented in [4]. Different UWB channel sounding techniques are discussed and approaches for the modelling of the UWB channel are reviewed.

A considerable work has been performed in [5–7] to characterize communication channels for general wireless applications. As MIMO systems operate at an unprecedented level of complexity to exploit the channel space-time resources, a new level of understanding of the channel space-time characteristics is required to assess the potential performance of practical multi-antenna links.

Empirical investigation of spatial correlation in UWB indoor channels has been presented in [8]. It was observed that the coherence distance falls with channel bandwidth in end-fire arrays but not in broadside arrays. The complex correlation decays with respect to distance in broadside arrays. It has been considered especially in line-of-sight communication. Strong dependence of spatial correlation as well as coherence distance on the channel centre frequency was observed in [8].

Spatial multiplexing single-input–multiple-output (SM-SIMO) UWB communication system using the TR technique has been proposed. The system with only one transmit antenna, using a spatial multiplexing scheme, can transmit several independent data streams to achieve a very high data rate. TR can mitigate not only the ISI but also the MSI caused by multiplexing the data streams simultaneously as in [9]. Antenna selection scheme for MIMO UWB communication system with TR is investigated in [10].

Time reversal technique has advantage in highly scattering environments to achieve signal focusing through transmitter-side processing that enables the use of simple receivers. The authors have also demonstrated UWB time reversal system architecture taking into account some practical constraints [11].

4.3 Methodology

In this work, it has been presented the measurements of a MIMO system under line-of-sight conditions. As it emphasizes on correlation based method, the design of the model and its mitigation techniques are explained as follows.

4.3.1 Spatial Correlation

Though the space-time focusing feature is one of the benefits, spatial multiplexing has a major role in MIMO systems. Without the expansion of bandwidth, high data rate can be achieved by using spatial multiplexing scheme with multiple transmit and receive antennas. In case of multipath channel, correlation is a critical factor in MIMO system for its performance and that is evaluated. It is mainly caused by inadequate antenna spacing in both transmitting and receiving side. It causes correlation between the received signals, which degrades the signal quality, capacity and bit error rate (BER) performance. Capacity increases and BER performance also increases as signal correlation decreases. The fading correlation between the array elements should be moderately low for a MIMO system for enhancement of performance.

The MIMO channel matrix is assumed to be independent of each other.

Other assumptions that can be made for such model analysis is as follows:

(i) The correlation between the receiving antennas is independent of the correlation between the transmitting antennas.
(ii) The effect of antenna coupling is neglected.
(iii) The transmitting and receiving correlation matrices are fixed.

For MIMO-UWB channel model design, the fixed correlation matrices have been included. Such model is known as Kronecker model [12–14], and it can be represented as,

$$H = R_{rx}^{1/2} H_W R_{tx}^{1/2} \qquad (4.1)$$

where,

H_w = channel matrix of independent channel realization.
$R_{tx}^{1/2}$ = transmit correlation matrix with dimension $M \times M$.
$R_{rx}^{1/2}$ = receive correlation matrix with dimension $N \times N$.
H = correlated channel.

4.3.2 Methods for Computing Spatial Correlation

1. Gathering a large amount of data in the target propagation environment. For that it becomes necessary to estimate a large number of correlation coefficients in an $M \times N$ MIMO system. For such MIMO system, consider MN spatial sub-channels. By correlating each pair would give rise to $(MN)^2$ correlation values. The major disadvantage of this approach is that it is most time-consuming, because of the estimation of large number of correlation coefficients.
2. Using fixed correlation matrices as:

    (i) Transmit correlation matrix
    (ii) Receive correlation matrix

    The transmit correlation matrix is given as:

$$R_{tx} = \begin{bmatrix} 1 & \rho_{Tx} & \rho_{Tx}^2 & \cdots & \rho_{Tx}^{M-1} \\ \rho_{Tx} & 1 & \rho_{Tx} & \cdots & \rho_{Tx}^{M-2} \\ \vdots & \vdots & \vdots & \ddots & \vdots \\ \rho_{Tx}^{M-1} & \rho_{Tx}^{M-2} & \rho_{Tx}^{M-3} & \cdots & 1 \end{bmatrix} \qquad (4.2)$$

Similarly, receive correlation matrix is given as:

$$R_{rx} = \begin{bmatrix} 1 & \rho_{Rx} & \rho_{Rx}^2 & \cdots & \rho_{Rx}^{N-1} \\ \rho_{Rx} & 1 & \rho_{Rx} & \cdots & \rho_{Rx}^{N-2} \\ \vdots & \vdots & \vdots & \ddots & \vdots \\ \rho_{Rx}^{N-1} & \rho_{Rx}^{N-2} & \rho_{Rx}^{N-3} & \cdots & 1 \end{bmatrix} \qquad (4.3)$$

IEEE 802.15.3a standard uses the fixed correlation model because of its simplicity. Also such model is used as a standardized one for the researchers.

For specific environmental condition, these correlation matrices R_{tx} and R_{rx} are appropriate and can be determined by selecting the values of transmit correlation coefficient, ρ_{Tx}, and receive correlation coefficient, ρ_{Rx}. It achieves a good result and a close match for the BER results of indoor channel model. The correlation

Table 4.1 Channel model parameter of IEEE 802.15.3a standard

Parameters	Specific values considered
Channel model	CM1 for line-of-sight communication
Frequency	3 GHz
Channel	Rayleigh fading channel
Modulation	QPSK
τ	5.05 ns

coefficient values ranges from 0 to 1. We present BER results for various MIMO UWB systems for various values of the channel correlation coefficient.

4.3.3 Proposed Method for Reduction of Correlation Effect

4.3.3.1 Virtual MIMO-UWB-Time Reversal Technique System

The time reversal (TR) technique is originated from under-water acoustics and ultrasonic. But now it has been used in many applications such as wireless communications, UWB communication as well as in other fields [15–19]. In that case TR technique was implemented as a filter at the transmitter side. By the help of the pre-filter, the system with only one transmit antenna can able to deliver several independent data streams at the same time. The spatial correlation is investigated, where a constant spatial correlation model with line-of-sight (LOS) channels for MIMO UWB has been applied. It has been referred to channel model1 (CM1) in the IEEE 802.15.3a standard [20]. The BER results using such method with an appropriate value of coefficient are shown as matching with indoor channel environment. It has been explained as follows.

The CIR as in [15] between the transmit antenna j and the receive antenna i is,

$$h_{i,j}(t) = \alpha^{i,j}\delta(t - \tau^{i,j}) \qquad (4.4)$$

where, α is the amplitude

τ is the delay and the value is considered for it as IEEE 802.15.3a standard from the Table 4.1. The TR matrix $\overline{H_{i,j}}(t)$ is used instead of $H_W(t)$ in Eq. 4.1, to calculate the BER performance of the MIMO UWB system.

$$H_{i,j}(t) = \begin{pmatrix} h_{1,1}(t) & h_{1,2}(t) & \cdots & h_{1,M}(t) \\ h_{2,1}(t) & h_{2,2}(t) & \cdots & h_{2,M}(t) \\ \vdots & \vdots & \ddots & \vdots \\ h_{N,1}(t) & h_{N,2}(t) & \cdots & h_{N,M}(t) \end{pmatrix} \qquad (4.5)$$

The matrix of the MIMO system as filter is given by:

$$\overline{H_{i,j}}(t) = \begin{pmatrix} \overline{h_{1,1}}(-t) & \overline{h_{2,1}}(-t) & \cdots & \overline{h_{N,1}}(-t) \\ \overline{h_{1,2}}(-t) & \overline{h_{2,2}}(-t) & \cdots & \overline{h_{N,2}}(-t) \\ \vdots & \vdots & \ddots & \vdots \\ \overline{h_{1,M}}(-t) & \overline{h_{2,M}}(-t) & \cdots & \overline{h_{N,M}}(-t) \end{pmatrix} \quad (4.6)$$

where,

$$\overline{h}(t) = h(t) \otimes h(-t) \quad (4.7)$$

The following parameters have been considered for evaluation of UWB channel model. It has been tested according to the IEEE standard 802.15.3a UWB model.

It is shown the virtual MIMO outperforms the true MIMO system in term of the BER performance. Another method to reduce the effect of correlation has been chosen by taking the Eigen value of the channel matrix for the computation of the system performance.

4.3.3.2 Eigen Values of the Correlation Matrix

The sub-channel correlation, power gains of supported Eigen modes, and branch power ratios are analyzed. The mutual information capacity is found to scale almost linearly with the MIMO array size, with very low variance. Eigen value of the correlation matrix is considered, that can be expressed as the relation:

$$\det(\lambda I - A) = 0 \quad (4.8)$$

where,

λ = the Eigen value of A and A = square matrix

4.4 Results and Discussion

Figure 4.1 shows the capacity results for $M = N = 2$, i.e., (2×2) for the systems operating in the CM1 with correlation coefficients ρ_{Tx} and ρ_{Rx} to be 0, 0.3 and 0.9 in the measured UWB LOS channel.

The capacity for different correlation factors is tested and it has been found that the capacity decreases with increase in the correlation factors and also decreases with increase in the number of correlated antenna elements. BER results for $M = N = 2$, that is (2×2) for the systems operating in the CM1 with correlation coefficients ρ_{Tx} and ρ_{Rx} to be 0, 0.3 and 0.9 and without applying time reversal in the measured UWB-LOS channel is shown in Fig. 4.2. It has been observed that the BER performance decreases with increase in the value of the correlation coefficients.

4 Performance Analysis of Correlated Channel for UWB-MIMO System

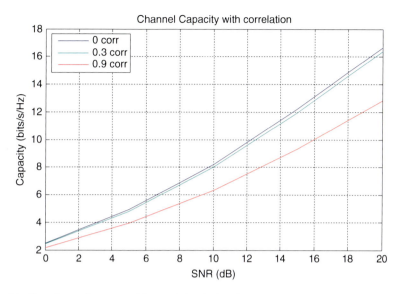

Fig. 4.1 Capacity performance with correlation in 2 × 2 systems

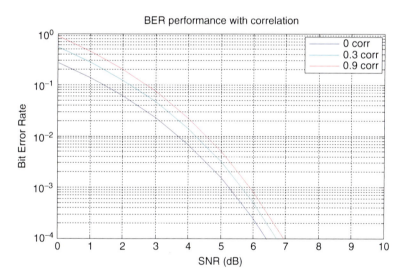

Fig. 4.2 BER performance with correlation

The BER results for $M = 1, N = 2$, that is (1×2) for the systems operating in the CM1 with correlation coefficients ρ_{Tx} and ρ_{Rx} to be 0, 0.3 and 0.9 is shown in Fig. 4.3, by applying virtual MIMO time reversal in the measured UWB -LOS channel. Also it has been shown that the BER performance is better than the one without time reversal.

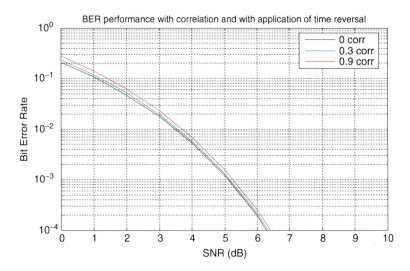

Fig. 4.3 BER performance with correlation and time reversal technique

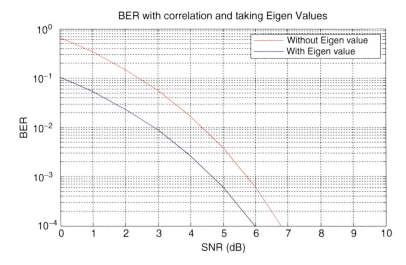

Fig. 4.4 BER with Eigen values

In Fig. 4.4, BER performance of 2 × 2 MIMO-UWB systems in the LOS indoor CM1 with correlation coefficients $\rho_{Tx} = \rho_{Rx} = 0.4$ is shown. The comparison of the BER shows its efficacy. It is considered with correlation and the Eigen value of the correlation matrix in shown. Here, it has been observed that, the BER performance is even better for Eigen values of the correlation matrix if considered.

4.5 Conclusion

The performance has been analyzed in terms of BER, in this paper. As the correlated MIMO model is considered, we have proposed different methods to reduce the impact of correlation in the MIMO-UWB channel. Also a comprehensive comparative analysis is done with a distance independent spatial correlation model. Comparison between the BER performance of a system with correlation using virtual MIMO time reversal technique and without using time reversal technique is evaluated. BER performance of the system with Eigen values and without Eigen values of the correlation matrix respectively are also evaluated. Capacity result for 2 × 2 systems with different correlation coefficient values are shown as the proof. These advantages include extended range, improved reliability in fading environments and higher data throughputs. Various models can be designed for MIMO-UWB application and also the correlation coefficients may be observed. Simultaneously, the array antenna may be considered for such application purpose, where the correlation factor may be studied, evaluated and kept as the future work.

References

1. A.F. Molisch, J.R. Foerster, Channel models for ultrawideband personal area networks. IEEE Wirel. Commun. **10**, 14–21 (2003)
2. W.Q. Malik, D.J. Edwards, Measured MIMO capacity and diversity gain with spatial and polar arrays in ultrawideband channels. IEEE Trans. Commun. **55**, 2361–2370 (2007)
3. T.K. Nguyen, H. Nguyen, F. Zheng, T. Kaiser, Spatial correlation in SM-MIMO-UWB systems using a pre-equalizer and pre-Rake filter, in *Proceedings of the IEEE International Conference on Ultra-Wideband (ICUWB)*, Nanjing, September 20–23, 2010, pp. 540–543
4. Z. Irahhauten, H. Nikookar, G.J.M. Janssen, An overview of ultra wide band indoor channel measurements and modeling. IEEE Microw. Wirel. Compon. Lett. **14**, 386–389 (2004)
5. M.N. Mohanty, S. Mishra, Design of MCM based wireless system using wavelet packet network and its PAPR analysis, in *IEEE Conference, ICCPCT*, Noorul Islam University, Kanyakumari, 2013, pp. 821–824
6. M.N. Mohanty, L.P. Mishra, S.K. Mohanty, Design of MIMO space-time code for high data rate wireless communication. Int. J. Comput. Sci. Eng. **3**(2), 693–696 (2011)
7. M.A. Jensen, J.W. Wallace, MIMO wireless channel modeling and experimental characterization, in *Space-time processing for MIMO communications*, ed. by A.B. Gershman, N.D. Sidiropoulos (Wiley, Chichester, 2005), pp. 1–39
8. W.Q. Malik, Spatial correlation in ultrawideband channels. IEEE Trans. Wirel. Commun. **7**, 604–610 (2008)
9. H. Nguyen, Z. Zhao, F. Zheng, T. Kaiser, Preequalizer design for spatial multiplexing SIMO-UWB TR systems. IEEE Trans. Vehic. Technol. **59**, 3798–3805 (2010)
10. H. Nguyen, F. Zheng, T. Kaiser, Antenna selection for time reversal MIMO UWB systems. IEEE Trans. Signal Process. 1–5 (2010)
11. N. Guo, B.M. Sadler, R.C. Qiu, Reduced-complexity UWB time-reversal techniques and experimental results. IEEE Trans. Wirel. Commun. **6**, 1–6 (2007)
12. T. Kaiser, F. Zheng, E. Dimitrov, An overview of ultrawide-band systems with MIMO. Proc. IEEE **97**, 285–312 (2009)

13. A. Paulraj, R. Nabar, D. Gore, *Introduction to space-time wireless communications* (Cambridge University Press, Cambridge, 2003)
14. S.L. Loyka, Channel capacity of MIMO architecture using the exponential correlation matrix. IEEE Commun. Lett. **5**, 369–371 (2001)
15. H. Nguyen, V.D. Nguyen, T.K. Nguyen, K. Maichalernnukul, F. Zheng, T. Kaiser, On the performance of the time reversal SM-MIMO-UWB system on correlated channels. Int. J. Antennas Propag. **2012**, 1–8 (2012)
16. F. Foroozan, A. Asif, Time reversal based active array source localization. IEEE Trans. Signal Process. **59**(6), 2655–2668 (2011)
17. B. Wang, Y. Wu, F. Han, Y.H. Yang, K.J.R. Liu, Green wireless communications: a time-reversal paradigm. IEEE J. Sel. Areas Commun. **29**(8), 1698–1710 (2011)
18. T. Shimura, H. Ochi, Y. Watanabe, T. Hattori, Time reversal communication in deep ocean – results of recent experiments, in *Proceedings of the IEEE Symposium on Underwater Technology (UT) and Workshop on Scientific Use of Submarine Cables and Related Technologies*, Tokyo, 5–8 Apr 2011
19. B. Van Damme, K. Van Den Abeele, O. Bou Matar, The vibration dipole: a time reversed acoustics scheme for the experimental localisation of surface breaking cracks. Appl. Phys. Lett. **100**(8), Article ID 084103, 3 pages (2012)
20. J. Foerster, Channel modeling sub-committee report final, IEEE, Document IEEE P802.15- 02/490r1-SG3a (2003), http://www.ieee802.org/15/pub/2003/Mar03/02490r1P802-15SG 3a-Channel-Modeling-Subcommittee-Report-Final.zip

Part II
Network Architectures, Protocols and Routing

Chapter 5
Node Failure Time Analysis for Maximum Stability Versus Minimum Distance Spanning Tree Based Data Gathering in Mobile Sensor Networks

Natarajan Meghanathan

Abstract A mobile sensor network is a wireless network of sensor nodes that move arbitrarily. In this paper, we explore the use of a maximum stability spanning tree-based data gathering (Max.Stability-DG) algorithm and a minimum-distance spanning tree-based data gathering (MST-DG) algorithm for mobile sensor networks. We analyze the impact of these two algorithms on the node failure times, specifically with respect to the node lifetime (the time of first node failure) and network lifetime (the time of disconnection of the network of live sensor nodes due to one or more node failures). Both the Max.Stability-DG and MST-DG algorithms are based on a greedy strategy of determining a data gathering tree when one is needed and using that tree as long as it exists. The Max.Stability-DG algorithm assumes the availability of the complete knowledge of future topology changes and determines a data gathering tree whose corresponding spanning tree would exist for the longest time since the current time instant; whereas, the MST-DG algorithm determines a data gathering tree whose corresponding spanning tree is the minimum distance tree at the current time instant. We observe a node lifetime – network lifetime tradeoff: the Max.Stability-DG trees incur a lower node lifetime due to repeated use of a data gathering tree for a longer time; on the other hand, the Max.Stability-DG trees incur a longer network lifetime.

5.1 Introduction

A mobile sensor network is a dynamically changing wireless distributed system of arbitrarily moving sensor nodes that operate under limited battery charge, memory and processing capacity. In addition, the bandwidth of these networks is also limited as well as the transmission range of the nodes is restricted to conserve the

---

N. Meghanathan (✉)
Department of Computer Science, Jackson State University, Jackson, MS, USA
e-mail: natarajan.meghanathan@jsums.edu

battery charge and to reduce collisions. With all of the above operating constraints, it is not a practically feasible solution to expect each of these sensor nodes to individually transmit their data (directly or through multi-hop route) to the far away control center, commonly called sink. In this context, several data gathering algorithms that focus on aggregating data from the individual sensor nodes through the use of a communication topology (like chain [1], cluster [2], tree [3], connected dominating set [4], and etc.) have been proposed. However, most of the research on data gathering algorithms for sensor networks has been for stationary environments where the sensor nodes are considered fixed at a particular location for the entire lifetime.

The common objective of many of the data gathering algorithms for the static sensor networks has been to conserve energy and maximize the node lifetime and network lifetime. In this context, in a recent research [5], we evaluated the performance of the data gathering algorithms based on different communication topologies and observed the minimum distance-spanning tree based data gathering (MST-DG) trees to be the most energy-efficient. However, with mobility, the network topology changes dynamically with time and thus, there is a need to determine stable data gathering trees that do not break frequently.

The first half of the paper proposes the maximum stability data gathering (Max. Stability-DG) algorithm, a benchmarking algorithm for the optimal number of tree discoveries in mobile sensor networks. Under the assumption that the entire knowledge of future topology changes is known, the algorithm operates according to the following greedy principle: Whenever a data gathering tree is required at time instant t, choose the longest-living data gathering tree from t. Such a strategy is continued for the rest of the data gathering session. The sequence of such longest-living data gathering trees incurs the minimum number of tree discoveries. The worst-case run-time complexity of the Max.Stability-DG tree algorithm is $O(n^2 T\log n)$ and $O(n^3 T\log n)$ when operated under sufficient-energy and energy-constrained scenarios respectively, where n is the number of nodes in the network and T is the total number of rounds of data gathering; $O(n^2\log n)$ is the worst-case run-time complexity of the minimum-weight spanning tree algorithm (we use Prim's algorithm [6]) used to determine the underlying spanning trees from which the data gathering trees are derived. A similar approach is adopted to determine the sequence of MST-DG trees – with the only difference being that the underlying spanning tree is a minimum distance spanning tree determined based on the local network topology and not at the future topology changes. In the second half of the paper, we conduct an exhaustive simulation study of the Max.Stability-DG trees vs. the MST-DG trees and analyze their impact on the node lifetime and network lifetime. The rest of the paper is organized as follows: Section 5.2 presents the algorithms to determine the Max.Stability-DG trees and MST-DG trees. Section 5.3 presents the simulation environment used and the performance metrics. Section 5.4 describes the simulation results observed for the node and network lifetimes. Section 5.5 concludes the paper.

5.2 Data Gathering Algorithms Based on Maximum Stability and Minimum Distance Spanning Trees

The system model adopted in this research is as follows: Each sensor node is assumed to operate with an identical and fixed *transmission range*. For the purpose of calculating the coverage loss, we also use the *sensing range* of a sensor node, considered in this research, as half the transmission range of the node. Basically, a sensor node can monitor and collect data at locations within the radius of its sensing range and transmit them to nodes within the radius of its transmission range. For coverage to imply connectivity, the transmission range per node has to be at least twice the sensing range of the nodes [7]. Data gathering proceeds in rounds. During a round of data gathering, data gets aggregated starting from the leaf nodes of the tree and propagates all the way to the leader node. An intermediate node in the tree collects the aggregated data from its immediate child nodes and further aggregates with its own data before forwarding to its immediate parent node in the tree.

We use the notions of static graphs and mobile graphs (adapted from [8]) to capture the sequence of topological changes in the network and determine a stable data gathering tree that spans over several time instants. A ***static graph*** is a unit-disk graph [9], representing a snapshot of the network at a particular time instant, wherein there exists an edge between any two nodes only if the physical distance between the two nodes is within in the transmission range of the sensor nodes. For any edge in a static graph, the weight of the edge is the Euclidean distance between the nodes constituting the two ends of the edge. The Euclidean distance for a link $i - j$ between two nodes i and j, currently at (X_i, Y_i) and (X_j, Y_j) is given by: $\sqrt{(X_i - X_j)^2 + (Y_i - Y_j)^2}$.

A ***mobile graph*** $G(i, j) = G_i \cap G_{i+1} \cap \ldots G_j; 1 \leq i \leq j \leq T$ (T is the total number of rounds of data gathering, corresponding to the network lifetime) is a logical graph that captures the presence or absence of edges in the individual static graphs. In this paper, we generate the sequence of static graphs by periodically sampling the network topology for every data gathering round. The weight of an edge in the mobile graph $G(i, j)$ is the geometric mean of the weights of the edge in the individual static graphs spanning $G_i, \ldots, G_j$. Since there exist an edge in a mobile graph if and only if the edge exists in the corresponding individual static graphs, the geometric mean of these Euclidean distances would also be within the transmission range of the two end nodes for the entire duration spanned by the mobile graph. Note that at any time, a mobile graph includes only ***live sensor nodes***, nodes that have positive available energy.

5.2.1 Maximum Stability Spanning Tree-Based Data Gathering (Max.Stability-DG) Algorithm

The Max.Stability-DG algorithm is based on a greedy look-ahead principle and the intersection strategy of static graphs. When a mobile data gathering tree is required at a sampling time instant t_i, the strategy is to find a mobile graph $G(i, j) = G_i \cap G_{i+1} \cap \ldots G_j$ such that there exists a spanning tree in $G(i, j)$ and no spanning tree exists in $G(i, j + 1) = G_i \cap G_{i+1} \cap \ldots G_j \cap G_{j+1}$. We find such an epoch $t_i, \ldots, t_j$ as follows: Once a mobile graph $G(i, j)$ is constructed with the edges assigned the weights corresponding to the geometric mean of the weights in the constituent static graphs $G_i, G_{i+1}, \ldots, G_j$, we run the Prim's minimum-weight spanning tree algorithm on the mobile graph $G(i, j)$. A spanning tree exists in $G(i, j)$ if and only if it is connected. We repeat the above procedure until we reach a mobile graph $G(i, j + 1)$ in which no spanning tree exists and there existed a spanning tree in $G(i, j)$. It implies that a spanning tree basically existed in each of the static graphs $G_i, G_{i+1}, \ldots, G_j$ and we refer to it as the mobile spanning tree for the time instants $t_i, \ldots, t_j$. To obtain the corresponding mobile data gathering tree, we choose an arbitrary root node for this mobile spanning tree and run the Breadth First Search (BFS) algorithm on it starting from the root node. The direction of the edges in the spanning tree and the parent-child relationships are set as we traverse its vertices using BFS. The resulting mobile data gathering tree with the chosen root node (as the leader node) is used for every round of data gathering spanning time instants $t_i, \ldots, t_j$. We then set $i = j + 1$ and repeat the above procedure to find a mobile spanning tree and its corresponding mobile data gathering tree that exists for the maximum amount of time since t_{j+1}. A sequence of such maximum lifetime (i.e., longest-living) mobile data gathering trees over the timescale T corresponding to the number of rounds of a data gathering session is referred to as the **Stable Mobile Data Gathering Tree**. Figure 5.1 presents the pseudo code of the Max.Stability-DG algorithm that takes as input the sequence of static graphs spanning the entire duration of the data gathering session.

While operating the algorithm under energy-constrained scenarios, one or more sensor nodes may die due to exhaustion of battery charge even though the underlying spanning tree may topologically exist. For example, if we have determined a data gathering tree spanning across time instants t_i to t_j using the above approach, and we come across a time instant t_k ($i \leq k \leq j$) at which a node in the tree fails, we simply restart the Max.Stability-DG algorithm starting from time instant t_k considering only the live sensor nodes (i.e., the sensor nodes that have positive available energy) and determine the longest-living data gathering tree that spans all the live sensor nodes since t_k. The *if* block segment in statement 8 of Fig. 5.1 handles node failures when run under energy-constrained scenarios.

Input: Sequence of static graphs $G_1, G_2, \ldots G_T$;
Total number of rounds of the data gathering session – T
Output: *Stable-Mobile-DG-Tree*
Auxiliary Variables: i, j
Initialization: $i = 1; j=1$; *Stable-Mobile-DG-Tree* = Φ
Begin *Max.Stability-DG Algorithm*
1   **while** ($i \leq T$) **do**
2   Find a mobile graph $G(i, j) = G_i \cap G_{i+1} \cap \ldots \cap G_j$ such that either $j = T$ or
        no spanning tree exists in $G(i, j+1)$ and there exists at least one in $G(i, j)$
3   *Mobile-Spanning-Tree*(i, j) = **Prim's Algorithm** ($G(i, j)$)
4   *Root*(i, j) = Choose a node randomly in $G(i, j)$
5   *Mobile-DG-Tree*(i, j) = **BFS** (*Mobile-Spanning-Tree*(i, j), *Root*(i, j))
6   *Stable-Mobile-DG-Tree* = *Stable-Mobile-DG-Tree* U { *Mobile-DG-Tree*(i, j) }
7   **for** each time instant $t_k \in \{t_i, t_{i+1}, \ldots, t_j\}$ **do**
            Use the *Mobile-DG-Tree*(i, j) in t_k
8   **if** node failure occurs at t_k **then**
                $j = k - 1$
                break
            end if
        end for
9       $i = j + 1$
10 **end while**
11 **return** *Stable-Mobile-DG-Tree*
End *Max.Stability-DG Algorithm*

Fig. 5.1 Pseudo code: maximum stability-based data gathering tree algorithm

5.2.2 Minimum Distance Spanning Tree-Based Data Gathering Algorithm

In our simulation studies, we compare the performance of the Max.Stability-DG trees with that of the minimum-distance spanning tree based data gathering (MST-DG) trees. The sequence of MST-DG trees for the duration of the data gathering session is generated as follows: If a MST-DG tree is not known for a particular round, we run the Prim's minimum-weight spanning tree algorithm on the static graph representing the snapshot of the network topology generated at the time instant corresponding to the round. Since the weights of the edges in a static graph represent the physical Euclidean distance between the constituent end nodes of the edges, the Prim's algorithm will return the minimum-distance spanning tree on the static graph. We then choose an arbitrary root node and run the BFS algorithm starting from this node. The MST-DG tree is the rooted form of the minimum-distance spanning tree with the chosen root node as the leader node. We continue to use the MST-DG tree as long as it exists. The leader node of the MST-DG tree remains the same until the tree breaks due to node mobility or node failures. When the MST-DG tree ceases to exist for a round, we repeat the above procedure. This way, we generate a sequence of MST-DG trees, referred to as the ***MST Mobile Data Gathering Tree***.

5.3 Simulation Environment and Performance Metrics

We conduct an exhaustive simulation study (in a discrete-event simulator developed by us in Java for mobile sensor networks) to compare the performance of the Max.Stability-DG and MST-DG trees under diverse conditions of network density and mobility. The MAC (medium access control) layer is assumed to be collision-free and considered an ideal channel (no interference). Sensor nodes are assumed to be both TDMA (Time Division Multiple Access) and CDMA (Code Division Multiple Access)-enabled [10]. Every upstream node (using a unique CDMA code) broadcasts a time TDMA schedule (for data gathering) to its immediate downstream nodes; a downstream node transmits its data to the upstream node according to this schedule.

The network dimension is 100 × 100 m. The number of nodes in the network is 100 and initially, the nodes are uniform-randomly distributed throughout the network. The sink is located at (50, 300), outside the network field. For a given simulation run, the transmission range per sensor node is fixed and is the same across all nodes. The network density is varied by varying the transmission range per sensor node of 25 m (representative of moderate density, with connectivity of 97 % and above) and 40 m (representative of high density, with 100 % connectivity).

Each node is supplied with limited initial energy (2 J per node) and the simulations are conducted until the network of live sensor nodes gets disconnected due to the failures of one or more nodes. The energy consumption model used is a first order radio model [11], according to which the energy lost by a node to run the transmitter or receiver circuitry and the transmitter amplifier are respectively: $E_{elec} = 50$ nJ/bit and $\in_{amp} = 100$ pJ/bit/m^2. We turn off the radios when a node does not intend to receive any transmissions. The energy lost in transmitting a message (of size k bits) over a distance d is given by: $E_{TX}(k, d) = E_{elec} * k + \in_{amp} * k * d^2$. The energy lost to receive a k-bit message is: $E_{RX}(k) = E_{elec} * k$.

We conduct constant-bit rate data gathering at the rate of 4 rounds per second (one round for every 0.25 s). The size of the data packet is 2,000 bits; the size of the control messages used for tree discoveries (network-wide flooding) is assumed to be 400 bits. Each sensor node will lose energy to transmit the 400-bit message over its entire transmission range and receive the message from each of its neighbor nodes. In high density networks, the energy lost due to receipt of the redundant copies of the control messages dominates the energy lost at a node for tree discovery.

The node mobility model used is the well-known Random Waypoint mobility model [12] with the maximum node velocity (v_{max}) being 3, 10 and 20 m/s

5 Node Failure Time Analysis for Maximum Stability Versus Minimum Distance... 61

Fig. 5.2 Average node and network lifetime (Transmission range = 25 m)

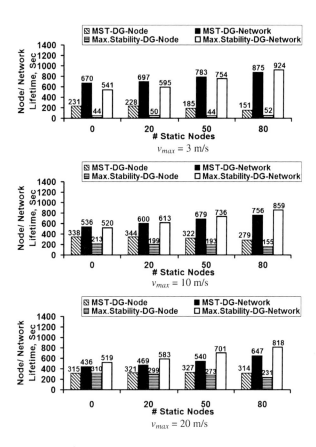

representing scenarios of low, moderate and high mobility respectively. Each node chooses a random target location to move with a velocity uniform-randomly chosen from $[0, \ldots, v_{max}]$, and after moving to the chosen destination location, the node continues to move by randomly choosing another new location and a new velocity. Each node continues to move like this, independent of the other nodes and also independent of its mobility history, until the end of the simulation. For a given v_{max} value, we also vary the dynamicity of the network by conducting the simulations with a variable number of static nodes (out of the 100 nodes) in the network. The values for the number of static nodes used are: 0 (all nodes are mobile), 20, 50 and 80.

We generated 200 mobility profiles of the network for a total duration of 6,000 s, for every combination of v_{max} and the number of static nodes. Every data point in the results presented in Figs. 5.2, 5.3, 5.4, 5.5 and 5.6 is averaged over these 200 mobility profiles. The performance metrics measured in the simulations are: (i) *Node Lifetime* – measured as the time at which the first node failure occurs due to depletion of battery charge. (ii) *Network Lifetime* – measured as the time of disconnection of the network of live sensor nodes, while the network would have

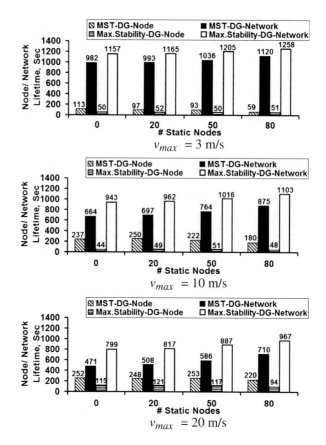

Fig. 5.3 Average node and network lifetime (Transmission range = 40 m)

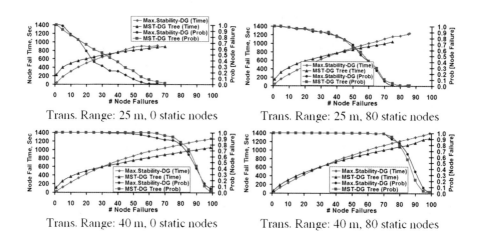

Fig. 5.4 Node failure times and probability of node failures [v_{max} = 3 m/s]

5 Node Failure Time Analysis for Maximum Stability Versus Minimum Distance... 63

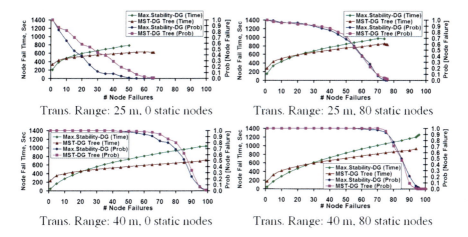

Fig. 5.5 Node failure times and probability of node failures [$v_{max} = 10$ m/s]

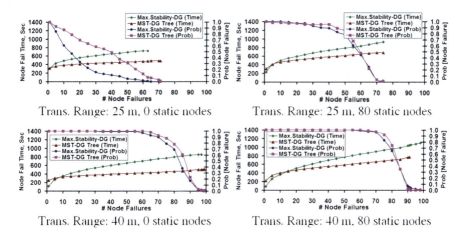

Fig. 5.6 Node failure times and probability of node failures [$v_{max} = 20$ m/s]

stayed connected if all the nodes were alive at that time instant. So, before confirming whether an observed time instant is the network lifetime (at which the network of live sensor nodes is noticed to be disconnected), we test for connectivity of the underlying network if all the sensor nodes were alive.

We obtain the distribution of node failures as follows: The probability for 'x' number of node failures (x from ranging from 1 to 100 as we have a total of 100 nodes in our network for all the simulations) for a given combination of the operating conditions (transmission range per node, v_{max} and number of static nodes) is measured as the number of mobility profile files that reported x number of node failures divided by 200, which is the total number of mobility profiles used for every combination of maximum node velocity and number of static nodes.

Similarly, we keep track of the time at which 'x' (x ranging from 1 to 100) number of node failures occurred in each of the 200 mobility profiles for a given combination of operating conditions and the values for the time of node failures reported in Figs. 5.4, 5.5 and 5.6 are an average of these data collected over all the mobility profile files.

5.4 Node Lifetime and Network Lifetime

We observe a tradeoff between node lifetime and network lifetime for maximum stability vs. minimum-distance spanning tree based data gathering in mobile sensor networks. The MST-DG trees incur larger node lifetimes (the time of first node failure) for all the 48 operating combinations of v_{max}, number of static nodes and transmission range per node. The Max.Stability-DG trees incur larger network lifetime for most of the operating conditions. Due to the unfairness in node usage resulting from the continued overuse of certain nodes as intermediate nodes (that receive aggregate data from one or more child nodes and transmit them to an upstream node in the tree) and leader node (that propagates the aggregated data to the sink), the Max.Stability-DG trees have been observed to yield a lower node lifetime, especially under operating conditions (like low and moderate node mobility with moderate and larger transmission range per node) that facilitate greater stability.

For the Max.Stability-DG trees, we observe an increase in node lifetime by as large as 200–400 % as we increase v_{max} from 3 to 10 m/s and operate the nodes at a moderate transmission range of 25 or 30 m. A further increase in v_{max} (i.e., from 10 m/s) to 20 m/s increases the node lifetime further by 50–100 %. A similar impact of node mobility on the node lifetime incurred with the MST-DG trees can also be observed, albeit at a lower percentage increase. The node lifetime for the MST-DG trees increases by about 50–100 % as we increase the maximum node velocity from 3 to 10 m/s. However, a further increase in the maximum node velocity from 10 to 20 m/s does not create a similar positive impact on the node lifetime; we observe the node lifetime to further increase by only about 10–20 %, and in case of lower transmission ranges per node, we even observe a 5 % decrease in node lifetime.

The node lifetime incurred for the MST-DG trees can be larger than that of the Max.Stability-DG trees by as large as 400 % at low and moderate levels of node mobility and by as large as 135 % at higher levels of node mobility. For a given level of node mobility, the difference in the node lifetimes incurred for the MST-DG trees and Max.Stability-DG trees increases with increase in the transmission range per node (for a fixed number of static nodes) and either remain the same or slightly increase with increase in the number of static nodes (for a fixed transmission range per node). For a fixed level of node mobility, the node lifetime measured for the Max.Stability-DG trees decreases by about 30–40 % and further by another 50–60 % as we increase the transmission range per node from 25 to 30 m and further to 40 m respectively. The MST-DG trees also suffer a similar decrease in node lifetime with increase in the transmission range per node, albeit at a lower scale (attributed to their relatively lower stability). At larger transmission ranges per node, the data gathering trees are bound to be more stable, and the negative impact of this on node lifetime is significantly felt in the case of the Max.Stability-DG trees. For a given transmission range per node, the negative impact associated with the use of static nodes on node lifetime is increasingly observed at v_{max} values of 3 and 10 m/s. At $v_{max} = 20$ m/s, since the network topology changes dynamically, even the use of 80 static nodes is not likely to overuse certain nodes and result in their premature failures. The node lifetime incurred with MST-DG trees is more impacted with the use of static nodes at low node mobility scenarios (Fig. 5.4) and the node lifetime incurred with the Max.Stability-DG trees is more impacted with the use of static nodes at moderate and higher node mobility scenarios (Figs. 5.5 and 5.6).

The Max.Stability-DG trees compensate for the premature failures of certain nodes by incurring a lower energy loss per round and energy loss per node due to lower tree discoveries and shorter tree height with more even distribution of the number of child nodes per intermediate node. As the dynamicity of the network increases, the data gathering trees become less stable, and this helps to rotate the roles of the intermediate nodes and leader node among the nodes to increase the fairness of node usage. All of these save significantly more energy at the remaining nodes that withstand the initial set of failures. As a result, we observe the Max. Stability-DG trees to observe a significantly longer network lifetime compared to that of the MST-DG trees for most of the operating conditions.

The magnitude of the difference in the network lifetime between the Max. Stability-DG and MST-DG trees increases with increase in v_{max} and transmission range per node. At low, moderate and high levels of node mobility, the network lifetime incurred with the Max.Stability-DG trees can be larger than that of the MST-DG trees by about 5–20 %, 15–40 % and 20–60 % respectively, with the difference increasing with increase in the transmission range per node. Similar range of differences in the network lifetime can be observed for the two data gathering trees at transmission ranges per node of 25, 30 and 40 m, with the difference increasing as the maximum node velocity increases. For a given v_{max} and transmission range per node, the number of static nodes does not make a significant impact on the difference in the network lifetime incurred with the two data gathering trees at moderate transmission ranges per node of 25 and 30 m. However, at larger transmission ranges per node of 40 m, the difference in the network lifetime decreases by about 15–35 %. This could be attributed to the relatively high stability of the Max.Stability-DG trees when operated at larger transmission ranges per node in the presence of more static nodes.

The network lifetime incurred with the two data gathering trees increases with increase in the number of static nodes for a given value of v_{max} and transmission range per node. For a given level of node mobility, the network lifetime increases with increase in transmission range per node; however, for the relatively unstable MST-DG trees, the rate of increase decreases with increase in v_{max} (requiring frequent tree reconfigurations). During a network-wide flooding, all nodes in the network tend to lose energy, almost equally. The Max.Stability-DG trees maintain a steady increase in the network lifetime with increase in transmission range per node for all levels of node mobility. For a given number of static nodes and a fixed transmission range per node, the network lifetime incurred with the Max.Stability-DG trees and MST-DG trees decreases by about 30–50 % and 50–100 % respectively as we increase v_{max} from 3 to 20 m/s, attributed to the energy loss incurred due to frequent tree discoveries.

For a given transmission range per node, with the absolute values of the node lifetime increasing with increase in the maximum node velocity and the network lifetime decreasing with increase in the maximum node velocity, we observe the maximum increase in the absolute time of node failures to occur at low node mobility. This vindicates the impact of network-wide flooding based tree discoveries on energy consumption at the nodes. Since all nodes are likely to lose the same amount of energy with flooding, the more we conduct flooding, the larger is the network-wide energy consumption. As a result, node failures tend to occur more frequently when we conduct frequent flooding. Thus, even though operating the network at moderate and high levels of node mobility helps us to extend the time of first node failure, the subsequent node failures occur too soon after the first node failure. This could be justified with the observation of flat curves for the MST-DG trees with respect to the distribution of node failure times (in Figs. 5.4, 5.5 and 5.6). The distribution of node failure times is relatively steeper for the Max.Stability-DG trees. The unfair usage of nodes in the initial stages does help the Max.Stability-DG trees to prolong the network lifetime. Aided with node mobility, it is possible for certain energy-rich nodes (that might have been leaf nodes in an earlier data gathering tree) to keep the network connected for a longer time by serving as intermediate nodes, and the energy-deficient nodes serve as leaf nodes during the later rounds of data gathering.

The impact of mobility in prolonging node failure lifetimes could also be explained by the lower probability of node failure observed for the Max.Stability-DG trees vis-à-vis the MST-DG trees when there are 0 static nodes (the plots to the left in Figs. 5.4, 5.5 and 5.6). At 80 static nodes, the probability of node failures for the two data gathering trees is about the same and is higher than that observed when all nodes are mobile. This is due to the repeated overuse of the intermediate nodes and leader node on relatively more stable data gathering trees. Thus, with the use of static nodes, even though the network lifetime can be marginally increased (by about 10–70 %; the increase is larger at moderate transmission range per node and larger values of v_{max}), the probability of node failures to occur also increases.

In terms of the percentage difference in the values for the network lifetime and node lifetime incurred with the two data gathering trees, we observe the Max.Stability-DG trees to incur a significantly prolonged network lifetime, beyond the time of first node failure. For a given transmission range per node and maximum node velocity, we observe the difference between the node lifetime and network lifetime for the Max.Stability-DG trees to increase significantly with increase in the number of static nodes. This could be attributed to the reduction in the number of

5 Node Failure Time Analysis for Maximum Stability Versus Minimum Distance... 67

flooding-based tree discoveries. For a given level of node mobility, we observe the difference in the node lifetime and network lifetime for the Max.Stability-DG trees to increase with increase in the transmission range per node. This could be again attributed to the decrease in the number of network-wide flooding based tree discoveries when operated at larger transmission ranges per node. Relatively, the MST-DG trees incur a very minimal increase in the network lifetime compared to the node lifetime, especially when operated at higher levels of node mobility. The network lifetime incurred with the Max.Stability-DG trees could be larger than the node lifetime as low as by a factor of 1.7 and as large as by a factor of 23. On the other hand, the network lifetime incurred with the MST-DG trees could be larger than the node lifetime as low as by a factor of 1.4 and as large as by a factor of 5.7.

One can also observe from Figs. 5.4, 5.5 and 5.6 that the number of node failures that require for the node failure time incurred with the Max.Stability-DG trees to exceed that of the node failure time incurred with the MST-DG trees decreases with increase in maximum node mobility. This could be attributed to the premature node failure occurring for the Max.Stability-DG trees when operated under low node mobility scenarios, with the time of first node failure for the MST-DG tree being as large as 400 % more than the time of first node failure for the Max.Stability-DG tree. On the other hand, at high levels of node mobility, the time of first node failure incurred with the MST-DG trees is at most 100 % larger than that of the Max. Stability-DG trees. Hence, the node failure times incurred with the Max.Stability-DG trees could quickly exceed that of the MST-DG trees at higher levels of node mobility. At the same time, the probability for node failures to occur (relatively low at moderate transmission ranges per node, low and moderate levels of node mobility) with the Max.Stability-DG trees converges to that of the MST-DG trees when operated at higher levels of node mobility and with larger transmission ranges per node. For a given transmission range per node and v_{max} value, with more static nodes, the Max.Stability-DG trees incur relatively more node failures than that of the MST-DG trees.

5.5 Conclusions

The main contribution of this paper in the area of data gathering for mobile sensor networks is the identification of a node lifetime – network lifetime tradeoff. The MST-DG trees sustain a longer node lifetime (as large as 400 % more) compared to that of the maximum stability-based data gathering trees. On the other hand, the maximum stability-based data gathering trees incur a longer network lifetime (the time of disconnection of the network of live sensor nodes) that can be as large as 60 % more compared to the MST-DG trees. The MST-DG trees suffer from an avalanche of node failures after the first node failure and this could be attributed to the larger; but equal, energy consumption of nodes across the network. Hence, even though the first node failure occurs after a prolonged run, the consequent node failures occur quickly. On the other hand, the Max.Stability-DG trees, with the tendency to use a tree as long as it exists, burn out the energy supplies at selected nodes; but, the variations in node usage does help to increase the time between successive node failures. Aided with mobility, the leaf nodes that have not been used much in the earlier part of the network lifetime, are likely to serve as intermediate nodes in the later part of the network lifetime, and vice-versa. The results from this research can serve as a pointer for someone exploring to design data gathering algorithms for mobile sensor networks to maximize either node lifetime or network lifetime. The results of this research indicate that if the stability-based data gathering algorithms for mobile sensor networks are designed to be energy-aware, one can prolong the node lifetime (time of first node failure due to exhaustion of battery charge).

References

1. S. Lindsey, C. Raghavendra, K.M. Sivalingam, Data gathering algorithms in sensor networks using energy metrics. IEEE Trans. Parallel Distrib. Syst. **13**(9), 924–935 (2002)
2. W. Heinzelman, A. Chandrakasan, H. Balakarishnan, Energy-efficient communication protocols for wireless microsensor networks, in *Proceedings of the Hawaiian International Conference on Systems Science*, Maui, HI, USA, Jan 2000
3. N. Meghanathan, An algorithm to determine energy-aware maximal leaf nodes data gathering tree for wireless sensor networks. J. Theor. Appl. Inf. Tech. **15**(2), 96–107 (2010)
4. N. Meghanathan, A data gathering algorithm based on energy-aware connected dominating sets to minimize energy consumption and maximize node lifetime in wireless sensor networks. Int. J. Interdiscip. Telecommun. Netw. **2**(3), 1–17 (2010)
5. N. Meghanathan, A comprehensive review and performance analysis of data gathering algorithms for wireless sensor networks. Int. J. Interdiscip. Telecommun. Netw. **4**(2), 1–29 (2012)
6. T.H. Cormen, C.E. Leiserson, R.L. Rivest, C. Stein, *Introduction to algorithms*, 3rd edn. (MIT Press, Cambridge, MA, 2009)
7. H. Zhang, J.C. Hou, Maintaining sensing coverage and connectivity in large sensor networks. Wirel. Ad Hoc Sens. Netw. **1**(1–2), 89–123 (2005)
8. A. Farago, V.R. Syrotiuk, MERIT: a scalable approach for protocol assessment. Mobile Netw. Appl. **8**(5), 567–577 (2003)
9. F. Kuhn, T. Moscibroda, R. Wattenhofer, Unit disk graph approximation, in *Proceedings of the Workshop on Foundations of Mobile Computing*, Philadelphia, PA, USA, Oct 2004, pp. 17–23
10. A.J. Viterbi, *CDMA: principles of spread spectrum communication*, 1st edn. (Prentice Hall, Englewood Cliffs, 1995)
11. T.S. Rappaport, *Wireless communications: principles and practice*, 2nd edn. (Prentice Hall, Upper Saddle River, 2002)
12. C. Bettstetter, H. Hartenstein, X. Perez-Costa, Stochastic properties of the random-way point mobility model. Wirel. Netw. **10**(5), 555–567 (2004)

Chapter 6
Wireless Networks: An Instance of Tandem Discrete-Time Queues

Nikhil Singh and Ramavarapu S. Sreenivas

Abstract We model end-to-end flows in an ad-hoc wireless network using a tandem of finite-size, discrete-time queues, located at the nodes along the routes used by the flows, with appropriate restrictions that capture the first- and second-order interference constraints. In addition, we assume there are no capture effects, that is, there is at most one arrival into a queue at any discrete-time instant. The half-duplex nature of communication also supposes there cannot be a simultaneous arrival and departure from a discrete-time queue. These queues are characterized by the channel access probabilities of the node. If the objective is to bound the buffer overflow probability at each queue along a flow, we show that is not necessary to maintain separate queues for each flow that is routed through a node. We present simulation results to support our conclusions. This observation significantly eases the implementation of the distributed algorithm that enforces end-to-end proportional fairness subject to constraints on the buffer overflow probabilities (Singh N, Sreenivas R, Shanbhag U (2008) Enforcing end-to-end proportional fairness with bounded buffer overflow probabilities. Technical Report UILU-ENG-08-2211, Aug 2008, Coordinated Science Laboratory, University of Illinois at Urbana-Champaign, Urbana).

In this paper we consider an ad-hoc wireless network with half-duplex links [1] that carries several flows between various source-destination pairs under a slotted-time medium access control (MAC) protocol. We assume that each node in the network has a finite buffer assigned to each flow routed through it, and it is of interest to keep

---

N. Singh
Yahoo! Labs, Champaign, IL, 61820, USA
e-mail: nsingh1@yahoo-inc.com

R.S. Sreenivas (✉)
Coordinated Science Laboratory (CSL) & Industrial and Enterprise Systems Engineering,
University of Illinois at Urbana-Champaign, Champaign, IL, USA
e-mail: rsree@illinois.edu

the buffer overflow probabilities at each node below a pre-determined value, in addition to other objectives. Reference [2] uses a class of *queue back-pressure random access algorithms* (QBRA), where the actual queue-lengths of the flows are used to determine a node's channel access probabilities. In this distributed algorithm, a node uses the queue-length information in a neighborhood to determine its channel access probability to achieve proportionally fair rates and queue stability. This scheme also has the downside that nodes need to maintain separate queues for each flow routed through it.

For a network of finite-buffer nodes it is of interest to achieve fairness subject to constraints on the buffer overflow probabilities. In [3], the authors present a distributed flow-based access scheme for slotted-time protocols, that is proportionally fair while respecting the bounds on buffer overflow probabilities at each node. The results in this paper imply that packets arriving along different incoming flows into a node can be stored in a common (finite-size) buffer while they are waiting to be serviced, without any violation of the buffer overflow constraints. Since nodes have only a limited amount of available memory to store data packets, maintaining separate queues for each flow can result in significant loss of performance. Our main contributions are:

1. We interpret each flow in the above mentioned wireless network as a tandem of discrete-time queues with constraints that represent primary- and secondary-interference constraints. We also assume there are no capture effects, which means there can be at most one packet arrival at any discrete-time instant. The half-duplex nature of communication prevents the possibility of a simultaneous arrival and departure of packets into a queue. We present an expression for the buffer overflow probability for this class of discrete-time queues.
2. We show that if the objective is keep the buffer overflow probability at each node below a common bound, then it is not necessary to maintain separate queues for each flow routed through a node.
3. The above observation is used in an *ns2* implementation of the procedure outlined in [3], and we present simulation results showing the satisfactory performance in terms of fairness and QoS when nodes maintain a single (merged) instead of separate queue for each flow.

The rest of the paper is organized as follows. Section 6.1 presents the network and discrete-time queue models used in the paper. In Sect. 6.2, we show that if all flows in the network have the same bound on the buffer overflow, nodes only need to maintain a single queue. Section 6.3 contains the details of the experimental results verifying the observations of Sect. 6.2. Conclusions are provided in Sect. 6.4.

6 Wireless Networks: An Instance of Tandem Discrete-Time Queues

6.1 Modeling the Network as a Tandem of Discrete Time Queues

6.1.1 Wireless Network Model

We assume the following:

1. Time is divided into slots of equal duration.
2. A successful transmission in a time-slot implies collision free data transmission in that slot.
3. The transmitting nodes always have data packets to transmit (i.e. we do not consider the arrival rates of packets for different flows, and assume that all flows have packets to transmit at all times).
4. Nodes cannot transmit and receive packets at the same time.
5. The receipt of more than one packet within the same time-slot will result in a collision.
6. Nodes in the network have a separate buffer of fixed size assigned to each flow routed through it.
7. We also assume there is a unique route for each flow within the network (which would be the case if we used AODV [4] as the routing protocol, for example).

Under the above assumptions, nodes in the wireless network can be seen as discrete time queues with certain arrival and departure rates and a flow in the network passes through a tandem of discrete time queues.

6.1.2 Buffer Overflow Probability of a Tandem of Discrete-Time Queues

The analysis of Ref. [5] for discrete-time queues that permits simultaneous arrivals and departures of multiple packets at a discrete time instant can be modified to the case of a discrete-time queue where (1) at most one packet arrives at any discrete-time instant, and (2) the simultaneous arrival and departure of packets into/from a queue is not permitted, would result in the following – for a discrete-time queue of capacity M, with a packet arrival probability p_a, and a probability p_d ($p_d > p_a$) of a packet departure from a non-empty buffer, the probability of seeing i-many packets at any time-instant in the buffer in steady state is given by the expression

$$\left(\frac{1-\rho}{1-\rho^{M+1}}\right)\rho^i, \text{ where } \rho = \frac{p_a}{p_d} \tag{6.1}$$

For the unrestricted discrete-time queue, Ref. [5] also shows that the joint stationary state probability of a tandem of discrete-time queues is the product of the distributions of each queue taken independently with an arrival probability of p_a, which is the probability of packet arrival into the first queue. This result uses the time-reversibility of the underlying Markov-chain and the mutual independence of the simultaneous states of the buffers, That is, the results in [5] can be thought of as the discrete-time analog of Jackson's result [6] involving tandems of $M/M/1$ queues.

As mentioned earlier, there are restrictions that one would have to enforce for a discrete-time queue to represent a node in a wireless network. For instance, as a node cannot transmit and receive information at the same time, the simultaneous occurrence of an arrival and a departure from the discrete-time queue at the node cannot be permitted. Secondary interference constraints place additional restrictions on the set of simultaneous events that can occur among neighboring nodes. Reference [7] notes that even for the unrestricted version of the discrete-time queue of Ref. [5], it is intractable to use balance equations to get an expression for the joint stationary probability for tandems of discrete-time queues. The restrictions we impose on the queues are only going to make it harder to use balance equations to arrive at an expression for the joint stationary probability for tandems of queues. It is not hard to construct examples of tandems of restricted, discrete-time queues where it can be shown that the joint stationary probability does not have a product form. In spite of this, it is possible to characterize the marginal probability distribution of each queue in the tandem.

In the restricted, discrete-time queue at most one packet is permitted to arrive, or depart from a single queue of size M. Additionally, there cannot be a simultaneous arrival and departure of a packet from the queue. For such a queue, the analysis of Ref. [5], with appropriate changes, results in the same expression as Eq. 6.1 for the probability of seeing i packets in the buffer at any time-instant. The probability of the queue of size M is non-empty is given by the expression $\frac{1-\rho^M}{1-\rho^{M+1}} \times \rho$ where $\rho = \frac{p_a}{p_d}$ and since the probability of a packet departure from a non-empty queue is p_d, the probability of a packet-departure from the discrete-time queue is given by $\underbrace{\frac{1-\rho^M}{1-\rho^{M+1}}}_{\leq 1} \times \underbrace{\rho}_{=\frac{p_a}{p_d}} \times p_d < p_a$. That is, the output process of the queue is geometrically distributed with a parameter that is no greater than the input parameter p_a. This observation holds for a tandem of discrete-time queues. That is, the output process of each queue is geometrically distributed with a parameter that is no greater than that of the input to the first queue (i.e. p_a). This observation is used in establishing a bound on the buffer-overflow probabilities at each queue in a tandem of discrete-time queues in the following theorem.

Theorem 6.1.1 *Consider a tandem of n discrete-time queues, each with buffer-size M, where at any discrete-time instant the probability of a packet-arrival into the first queue is p_a, and the probability of a packet-departure from the i-th, non-empty queue is p_{di}, ($i = 1,2,\ldots,n$). If $p_{dj} = \min_{i=1,\ldots,n} \{p_{di}\}$, and $\rho_{max} \left(= \frac{p_a}{p_{dj}}\right)$*

$< \frac{M}{M+1}$, then, the probability of seeing M packets in the i-th queue ($i = 1 \ldots n$) is no greater than $\left(\frac{1-\rho_{max}}{1-\rho_{max}^{M+1}}\right)\rho_{max}^{M}$.

Proof We first note that $\left(\frac{1-\rho}{1-\rho^{M+1}}\right)\rho^{M}$, increases monotonically with respect to ρ if $\rho \leq \frac{M}{M+1}$. Let p_{ai} be the probability of a packet arrival into the i-th queue, we know $p_{ai} \leq p_a$. If $\rho_i = \frac{p_{ai}}{p_{di}}$, since $p_{di} \geq p_{dj}$, it follows that $\rho_i \leq \rho_{max} < \frac{M}{M+1}$. The observation follows directly from the monotonicity property mentioned above.

Let β be an acceptable upper-bound on the buffer overflow probability. A direct consequence of Theorem 6.1.1 is that if we are able to pick a p_a such that

$$\frac{p_a}{p_{dj}}(= \rho_{max}) < \left[\frac{\beta}{1+\beta}\right]^{1/M}, \qquad (6.2)$$

then the buffer overflow probability at the i-th queue in the tandem of discrete-time queues will be no higher than β at all queues.

In [3], this observation is used in a convex programming solution to the problem of enforcing proportional fairness in the presence of constraints on the buffer overflow probabilities, in ad hoc wireless networks. Typically optimization problems for slotted-time protocols involve selecting the optimal access probabilities for each node such that some performance function is maximized, subject to relevant constraints (for example, [2, 3]).

If a flow is routed through nodes $i \rightarrow j \rightarrow k$, then the probability of packet arrival into (departure from) the discrete-time queue at node j is the single attempt success probability of link (i, j) (link (j, k)). Generalizing this observation, each flow in the network can be modeled as a tandem of restricted, discrete-time queues, where the idiosyncratic constraints of ad-hoc wireless networks are embedded in the expressions for these single attempt link success probabilities (cf. Eq. 6.3, Sect. 6.2). For the present,[1] each node through which a flow is routed, maintains a restricted, discrete-time queue for that flow.

6.2 Single Queue vs Multiple Queues

In this section we show that if all the flows in the network have the same requirement on the bound of buffer overflow probability, the nodes in the network do not need to maintain separate queues for each flow.

---

[1] This can be changed with impunity following the justification in Sect. 6.2.

6.2.1 Link Success Probability Expression

An ad-hoc wireless network carrying a collection of flows, is represented as an undirected graph $G = (V, E)$, where V represents the set of *nodes*, and $E \subseteq V \times V$ is a symmetric relationship (i.e., $(i, j) \in E \Leftrightarrow (j, i) \in E$), that represents the set of bi-directional *links*. We assume all links of the network have the same capacity, which is normalized to unity. The 1-hop neighborhood of node $i \in V$ is represented by the symbol $Ni)$. When a node i communicates with a node $j \in Ni)$, we can represent it as an appropriate orientation of the link (i, j) in E, where i is the origin and j is the terminus. The context in which $(i, j) \in E$ is used should indicate if it is to be interpreted as a directed edge with i as origin and j as terminus. The set of flows, using a link $(i, j) \in E$ with i (j) as origin (terminus), is denoted by $\mathcal{F}(i, j)$.

When node i intends to transmit data to node $j \in Ni)$ for the l-th flow ($l \in \mathcal{F}(i, j)$), it would transmit data in the appropriate time-slot with probability $\tilde{p}_{i,j,l}$. $\tilde{P}_{i,j} = \sum_{l \in \mathcal{F}(i,j)} \tilde{p}_{i,j,l}$, denotes the probability that node i transmits data to node j, and $\tilde{P}_i = \sum_{j \in V} \tilde{P}_{i,j}$, denotes the probability that node i will be transmitting to some node in its 1-hop neighborhood for some flow. The probabilities $\tilde{p}_{i,j,l}$'s should be chosen such that $\tilde{P}_i$ is not greater than unity for any node $i \in V$.

The probability of successful data transmission over link $(i, j) \in E$ for flow $l \in \mathcal{F}(i, j)$, denoted by $\mathcal{S}_{i,j,l}$, is given by the expression

$$\mathcal{S}_{i,j,l} = \tilde{p}_{i,j,l} \times \left(1 - \sum_{(j,m) \in E, n \in \mathcal{F}(j,m)} \tilde{p}_{j,m,n}\right) \times \left\{\prod_{o \in N(j) - \{i\}} \left(1 - \sum_{(o,p) \in E, q \in \mathcal{F}(o,p)} \tilde{p}_{o,p,q}\right)\right\}. \tag{6.3}$$

Assuming there are separate queues for each flow at each node, the probability of a departure, p_{di}^l, from the discrete-time queue for the l-th flow in node i (assuming it has a packet to send) is given by the above expression. If i is the source of the l-th flow (i.e. there is always a packet to send at node i) the above expression is the probability of arrival, p_{a1}^l, into the first queue in the tandem (which is located at node j). The above equation also ensures the service-constraint $\sum_{j \in N(i), l \in \mathcal{F}(i,j)} p_{di}^l \leq 1$.

6.2.2 Merging Queues

Consider a wireless network where all flows in the network have the same requirement on the bound of the buffer overflow probability stated in the context of

Theorem 6.1.1. That is, for any flow l, $\rho^l_{max} \leq \rho_{max}$, where $\rho^l_{max} = \frac{p^l_{a1}}{p^l_{dj}}$, $p^l_{dj} = \min_{i=1,\ldots,n}\{p^l_{di}\}$, and p^l_{ai} (p^l_{di}) is the probability of a packet arrival (packet departure, conditioned on the queue at node i, that the l-th flow is routed through, is non-empty).

We assume that all the nodes in the network are accessing the channel with floor access probabilities so as to satisfy the buffer overflow bounds. If ρ_{max} meets the requirement of Eq. 6.2, following Theorem 6.1.1, we have the fact that the probability of a buffer overflow at any node, for any flow, is bounded above by β.

Let us consider a node A, with a total of n flows routed through it, which maintains separate queues for each flow. Additionally, at node A,

1. Let $\tilde{p}_l(\tilde{p}_{i,j,l}$ in Eq. 6.3) be the optimal floor access probability used by node A to transmit packets of flow l in any time slot.
2. $\hat{P}_l$ denotes the probability that the receiving node for flow l and all the one hop neighbors of the receiving node (except transmitting node) are silent in any time slot.
3. $l \in A$ implies that flow l routes through node A. The access intensity for any flow $l \in A$ is given by $\rho^l = p^l_{aA}/p^l_{dA} \leq \rho_{max}$, where $p^l_{dA} = \tilde{p}_l \hat{P}_l$ (cf. Eq. 6.3).
4. The overall access probability for node is $\tilde{P}_A = \sum_{l \in A} \tilde{p}_l$.

Now if n queues are merged into a single queue and for each packet in the queue, node A uses $\tilde{P}_A$ to access the channel, then,

1. The effective arrival rate of the single merged queue is given by $\sum_{l \in A} p^l_{aA}$.
2. The effective departure rate of the single combined queue is given by the expression $\left(\tilde{P}_A \frac{\sum_{l \in A} \hat{P}_l p^l_{aA}}{\sum_{l \in A} p^l_{aA}}\right)$.

Theorem 6.2.1 *The traffic intensity of the merged queue is at most ρ_{max}.*

Proof Let us compute the $\hat{\rho}$ for the merged queue at node A.

$$\hat{\rho} = \frac{\sum_{l \in A} p^l_{aA}}{\tilde{P}_A \left(\frac{\sum_{l \in A} \hat{P}_l p^l_{aA}}{\sum_{l \in A} p^l_{aA}}\right)} = \frac{\left(\sum_{l \in A} p^l_{aA}\right)^2}{\left(\sum_{l \in A} \tilde{p}_l\right)\left(\sum_{l \in A} \hat{P}_l p^l_{aA}\right)}$$

$$= \frac{\left(\sum_{l \in A} p^l_{aA}\right)^2}{\left(\sum_{l \in A} \left((p^l_{aA})^2/\rho^l\right)\right) + \sum_{l \in A} \tilde{p}_l \sum_{j \in A, j \neq l} \hat{P}_j p^l_{aA}}. \qquad (6.4)$$

Now, let us look at one term in the second summand in the denominator,

$$\tilde{p}_l \sum_{j \in A, j \neq l} \hat{P}_j p^l_{aA} = \tilde{p}_l \sum_{j \in A, j \neq l} \frac{p^j_{dA} p^l_{aA}}{\tilde{p}_j} = p^l_{aA} \sum_{j \in A, j \neq l} \frac{p^j_{aA} \tilde{p}_l}{\rho} \tilde{p}_j \frac{p^j_{dA}}{p^l_{dA}}.$$

Using this in (6.4), along with the observation that $\rho^l < \rho_{max}$ we have,

$$\hat{\rho} \le \frac{\rho_{max}(\sum_{l \in A} p_{aA}^l)^2}{(\sum_{l \in A} (p_{aA}^l)^2 + \sum_{l \in A} \sum_{j \in A, j \ne l} p_{aA}^l p_{aA}^j \frac{\tilde{p}_l}{\tilde{p}_j} \frac{p_{dA}^j}{p_{dA}^l})}$$
$$\le \frac{\rho_{max}(\sum_{l \in A} p_{aA}^l)^2}{(\sum_{l \in A} (p_{aA}^l)^2 + \sum_{l \in A} \sum_{j \in A, j} p_{aA}^l p_{aA}^j (\frac{\tilde{p}_l}{\tilde{p}_j} \frac{p_{dA}^j}{p_{dA}^l} + \frac{\tilde{p}_j}{\tilde{p}_l} \frac{p_{dA}^l}{p_{dA}^j}))}. \quad (6.5)$$

Let $x = \frac{\tilde{p}_l}{\tilde{p}_j}$ and $y = \frac{p_{dA}^l}{p_{dA}^j}$, using the fact that $x/y + y/x \ge 2$, we have, $\left(\frac{\tilde{p}_l}{\tilde{p}_j} \frac{p_{dA}^j}{p_{dA}^l} + \frac{\tilde{p}_j}{\tilde{p}_l} \frac{p_{dA}^l}{p_{dA}^j} \right) \ge 2$. Therefore,

$$\left(\sum_{l \in A} p_{aA}^l \right)^2 \le \left(\sum_{l \in A} (p_{aA}^l)^2 + \sum_{l \in A} \sum_{j \in A, j} p_{aA}^l p_{aA}^j \left(\frac{\tilde{p}_l}{\tilde{p}_j} \frac{p_{dA}^j}{p_{dA}^l} + \frac{\tilde{p}_j}{\tilde{p}_l} \frac{p_{dA}^l}{p_{dA}^j} \right) \right).$$

Hence, from (6.5), we get, $\hat{\rho} \le \rho_{max}$.

Using a single queue instead of multiple queue at each node has the advantage that nodes in the network do not need to maintain separate queue for each flow routed through it, to respect the buffer overflow bound for each flow.

6.3 Preliminary Results

In this section, we describe the experiments that show how ST-MAC protocol [8] performs when the RTS-signal in the corresponding RTS-slot is transmitted with a probability as determined by the optimization algorithm discussed in [3]. For this, we compare the performance of the ST-MAC protocol using different bounds on the flow's buffer overflow probabilities (i.e. different traffic intensities). These experiments involved the implementation of the optimization algorithm within the *ns2* network simulator [9]. A detailed implementation of the 802.11-MAC protocol already exists as a part of the simulation model.

The results presented in this section, provide an understanding of the performance of the optimization algorithm presented in [3], in terms of fairness and buffer overflow probabilities. It is important to point out that in *ns2* network simulator, in addition to transmitting data for unicast flows in the network, nodes also perform additional functions such as route discovery and maintenance using broadcast packets, for example, the broadcasting of route request queries. The simulation results demonstrate the performance of ST-MAC protocol in presence of these additional network traffic.

6.3.1 Traffic and Mobility Models Used in the Experiments

The pause-time is kept constant and equal to the simulation time (i.e. there is no mobility) in all our experiments. Lucent's WaveLAN with 2 Mbps bit rates and 250 m-transmission range is used for the radio model. An omnidirectional antenna is used by the mobile/wireless nodes. The carrier sense range is the same as the transmission range in our simulation experiments, i.e. the packets are assumed to interfere with each other only when a receiver is within the transmission range of two sources that are transmitting simultaneously. We chose AODV [4] as the ad hoc routing protocol. Each flow in the network has the same bound on the buffer overflow probability.

In the simulations the retry limit for data packets is set to 4 for 802.11-MAC, which is the default 802.11-MAC long-retry limit. In case of ST-MAC protocol, there is no retry limit, i.e., the nodes keep trying to transmit the packet until its successful. In simulations, at each node a single Drop Tail Queue of size 50 packets was selected as an interface queue for both these protocols.

We then choose ten random source-destination pairs in the network shown in Fig. 6.1a, simulated within a 1,000 × 1,000 m field. Each source generates data packets of 512 bytes each. We run 20 different simulation instances with this fixed network layout and source-destination pairs, by varying the start times of each flow in the network by some milliseconds. Depending on the start time of the flows, AODV chooses different routes for different simulation instances, which are not same in all the instances.

For our simulation comparisons, we compared the following for scenarios:

- Nodes using 802.11 as MAC protocol, with TCP connections between the source-destination pairs. This provides us with the reference frame, for comparisons using different performance metrics, among the optimization algorithms using different bounds on buffer overflow probabilities and step sizes, when nodes in the network use ST-MAC protocol and CBR traffic.
- ST-MAC protocol with constant-bit-rate (CBR) flows between each source-destination pair, ρ varying from 0. 86 to 1. 0 for each flow.

The simulation time for each simulation is 620 s. The nodes start using the flow rates and access probabilities given by optimization algorithm [3] after 120 s from the start of the simulation and the simulation continues for additional 500 s. The nodes keep updating the flow rates and access probabilities until the simulation terminates.

In the case of 802.11-MAC protocol, the TCP connections are started when the simulation starts. For ST-MAC, however, we wait a total of 120 s before the nodes start transmitting at optimal rates. During this time, the sources transmit at very low rates so as to prevent any network congestion.

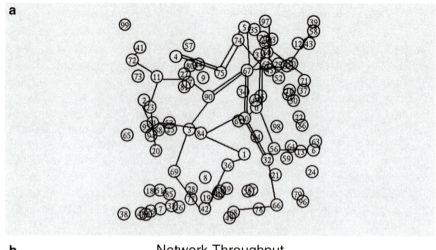

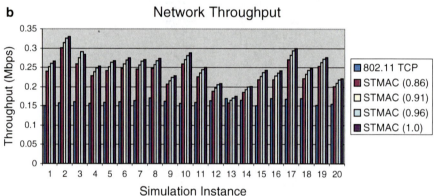

Fig. 6.1 Simulated network and throughput. (**a**) Ad hoc wireless network with 100 nodes, 10 connections. (**b**) Average network throughput

6.3.2 Results

Figures 6.1b and 6.2a, b show the plots of the average network throughput, packet-delivery ratios and end-to-end delays for different simulation runs of the scenarios mentioned in the earlier section. In case of 802.11-MAC the average packet delivery ratio and end to end delay is low and the because the packets are dropped by 802.11-MAC after four retries. This also contributes to lower the throughput in case of TCP.

In case of ST-MAC protocol, as ρ increases the network throughput increases but at a cost of larger delays and higher packet loss. As the value of ρ is increased, the network delay increases because of increase in queuing delays. Also, we can observe that when $\rho = 1$, the packet delivery ratio decreases as more packets are getting dropped because of buffer overflows.

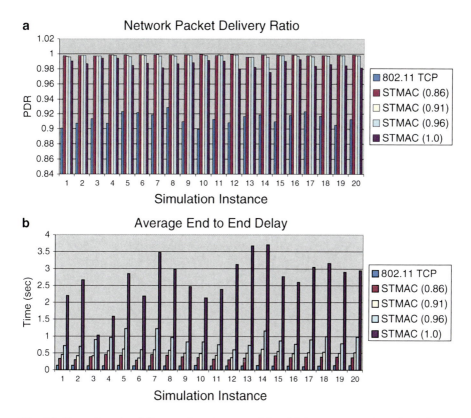

Fig. 6.2 Network packet delivery ratio and average delay. (**a**) Average network packet delivery ratio. (**b**) Average network end to end delay

6.4 Conclusion

Each end-to-end flow in an ad-hoc wireless network of half-duplex nodes using a slotted time protocol can be viewed as a tandem of discrete-time queues with restrictions that capture constraints specific to wireless networks. We derive a bound on the buffer overflow probabilities for each queue in the tandem. We then suppose that all queues, irrespective of the flows they serve, are subject to a common buffer overflow bound. We show that for this case there is no need to maintain separate queues for each flow that is routed through a host. These flows can all be merged into a single queue and served as if there were just one flow, and the buffer overflow bounds will still be met. This observation finds use in the implementation of the distributed algorithm that enforces proportional fairness subject to buffer overflow constraints outlined in Ref. [3]. We present some simulation results of this implementation.

References

1. M. Gast, *802.11 Wireless Networks: The Definitive Guide* (O'Reilly, Sebastapol, 2002)
2. J. Liu, A. Stolyar, Distributed queue length based algorithms for optimal end-to-end throughput allocation and stability in multi-hop random access networks, in *Proceedings of the 45th Annual Allerton Conference on Communication, Control, and Computing*, Monticello, Sept 2007
3. N. Singh, R. Sreenivas, U. Shanbhag, Enforcing end-to-end proportional fairness with bounded buffer overflow probabilities. Technical Report UILU-ENG-08-2211, Aug 2008, Coordinated Science Laboratory, University of Illinois at Urbana-Champaign, Urbana
4. C.E. Perkins, E.M. Royer, Ad-hoc on-demand distance vector routing, in *Proceedings of the 2nd IEEE Workshop on Mobile Computing Systems and Applications*, New Orleans, 1999, pp. 90–100
5. J. Hsu, P. Burke, Behavior of tandem buffers with geometric input and markovian output. IEEE Trans. Commun. **24**(3), 358–361 (1979)
6. R. Jackson, Queueing systems with phase type service. Oper. Res. **5**, 109–120 (1954)
7. K. Bharath-Kumar, Discrete-time queueing systems and their networks. IEEE Trans. Commun. **28**(2), 260–263 (1980)
8. N. Singh, R.S. Sreenivas, On distributed algorithms that enforce proportional fairness in ad-hoc wireless networks, in *Proceedings of the 5th International Symposium on Modeling and Optimization in Mobile, Ad Hoc, and Wireless Networks*, Limassol, 2007
9. The VINT Project, The Network Simulator – NS2, Report, 2004

Chapter 7
An Equal Share Ant Colony Optimization Algorithm for Job Shop Scheduling Adapted to Cloud Environments

Rajesh Chaukwale and Sowmya Kamath S

Abstract The problem of efficiently scheduling jobs on several machines is an important consideration for Cloud computing. Task scheduling in Cloud Environment is a recognised NP-hard problem and hence methods that focus on producing an exact solution can prove insufficient in finding an optimal resolution to JSSP. Hence, in such cases, heuristic methods can be employed to find a good solution within reasonable time. In this paper, we study the conventional ACO algorithm and propose two Load Balancing ACO algorithms for task scheduling in Cloud Environment. We also present the observed results, and discuss them with reference to the FCFS scheduling algorithm currently used. It is observed that the proposed algorithm gives better results for every problem size. Also the proposed algorithms are adapted and applied to Task scheduling in Cloud Environment and is found to give better results.

7.1 Introduction

There has been extensive research and development in the Cloud computing, the main focus of this development being user applications. The technology aims to provide distributed, virtualized resources to its end users. Since a user application may use thousands of virtual machines, it is difficult to manually assign tasks to such virtual machines. Clearly, there is a need for an efficient scheduling algorithm which can simplify this job of task scheduling in Cloud Environments.

The classical Job Shop Scheduling Problem (JSSP) considers the problem of efficiently scheduling a finite number of jobs to a finite number of machines for processing. Each job consists of a sequence of operations which have to be

---

R. Chaukwale • S. Kamath S (✉)
Department of Information Technology, National Institute of Technology Karnataka,
Surathkal, Mangalore 575025, India
e-mail: rajwithfriends@gmail.com; sowmyakamath@ieee.org

processed using a specified machine for a specified amount of time, without any interruption. The operations belonging to the same job have a technological sequence and none of them should begin processing before the preceding operation has finished its execution. The challenge is to find a feasible schedule consisting of the assignment of operations on machines without violating these constraints. Thus, JSSP is a NP-hard combinatorial optimization problem [1].

In this paper, we present an analysis of the application of the meta-heuristic Ant Colony Optimization technique for the task scheduling in Cloud environment. Ant Colony Optimization (ACO) technique is inspired by foraging behaviour of ants in nature. ACO tries to mimic the observed behaviour of ants while they conduct a search for an efficient path to follow to carry their food back to the nest. In a similar fashion, in ACO, the concept of an ant is considered. Each ant constructively builds a solution to the problem at hand by making decisions using path probabilities at each decision point.

ACO has been used extensively to present effective solutions to many combinatorial optimization problems like Travelling Salesman problem and Vehicle routing problem. In this paper, we apply and analyze the effectiveness of Ant colony optimization for Cloud environment. Furthermore, the results of ACO are improved by adding a load balancing factor while calculating probabilities. It was observed that, the inclusion of load balancing factor improves the results drastically and the proposed algorithm outperforms conventional ACO.

The paper is organized as follows – Sect. 7.2 presents a discussion of related work in the area. Section 7.3 sheds more light on the intricacies of the Ant colony optimization. Section 7.4 describes the proposed algorithms. Section 7.5 presents the results obtained and their analysis, followed by conclusion and references.

7.2 Related Work

ACO and Task scheduling problem in cloud has been a problem of constant research and has attracted the attention of several researchers. In recent years, researchers have proved that ACO performs well as compared to other meta-heuristic approaches when applied to scheduling problems [2, 3]. The first Ant algorithm was proposed by Dorigo et al. [4] in 1992 as a way to solve the Travelling Salesman Problem (TSP). The proposed Ant System [5] was improved to ACO. They suggested ACO based approaches to solve some of the complex quadratic assignment problems. When it comes to the field of scheduling, ACO has been successfully applied to solve challenging problems like vehicle routing, graph colouring, sequential routing, etc. Also, den Besten and Juan [6] applied ACO for single machine weighted tardiness problem which is also a NP-hard scheduling problem and consists of the problem of efficiently scheduling a given set of jobs on a single machine. Hui Yan, Xue-Qin Shen, Xing Li, Ming-Hui Wu [7, 8] applied ACO to task scheduling in Grid computing and proved that it performed better.

We conducted a study on how ACO can be applied to the problem of scheduling. It was found that ACO can be used to compete with the results of other meta-heuristic approaches like Genetic and Bee intelligence algorithms. Also, most of the ACO based algorithms proposed previously do not take the machine load into consideration. Hence, we propose an algorithm for Job Shop scheduling problem (JSSP) based on the concept of an equal-share factor. This factor helps in making sure that each machine gets equal amount of load while scheduling tasks. The same algorithm is extended for its application in cloud environment. As see from experimental results, the proposed algorithm outperformed conventional ACO.

7.3 Ant Colony Optimization (ACO)

An Ant system is based on the interesting foraging behaviour of ants observed in nature [4]. Ants are capable of finding the shortest path from food to their nest, without using visual cues. They actually use a hormone called pheromone deposited by other ants and its concentration levels to decide the best path. While walking, ants deposit pheromone on the path and the ants coming later choose the path with a greater concentration of the pheromone, since this indicates the way to the nest. The technique based on this idea is called Ant Colony Optimization.

The basic idea in ACO is to use a population of ants to iteratively build a solution by continually applying a decision policy based on probability until a solution is found. Ants that find a good solution mark their paths with pheromones. So, in the next iteration, ants are attracted towards to the pheromone which results in greater chances of following good paths. ACO has a memory which stores the components of the path being traversed. So, at the end of any iteration we can get the path used by the ants.

In-order to apply ACO algorithm, the optimization problem must be plotted in a graphical representation G. The pheromone level at each edge is initialized to a positive real value c. Each ant is positioned at the starting node. The ant then starts traversing the graph using the values obtained from probabilistic computations, until it reaches the destination node. The path used by the ant is recorded and the best solution will be recorded. The pheromone amount of the path used by ant is then determined and another ant will start its traversal if the stopping criteria is not met [9].

The ant uses a probabilistic computation to determine the next path during its traversal. The computation is based on two parameters: the pheromone present along the edge and the weight of the edge. The probability to move from node i to node j for each kth ant at time t can be defined as:

$$P(i,j) = \begin{cases} \dfrac{\tau(i,j)^\alpha * \eta(i,j)^\beta}{\sum_{u \in S_k} \tau(i,u)^\alpha * \eta(i,u)^\beta}, & \text{if } j \in S_k \\ 0, & \text{otherwise} \end{cases} \quad (7.1)$$

```
begin
    Initialize the parameters
    While (Stopping criterion not met)
        Position each ant at Start node
        while(Stop when each ant has build a solution)
            for each ant do
                Choose next node by pheromone trail & edge weight
            end for
        end while
        Update the pheromone
    end while
end
```

Fig. 7.1 Procedure for ACO

where $\tau(i, j)$ represents the pheromone trail, $\eta(i, j)$ represents the edge weight. The parameters α and β determine the degree to which pheromone level is used as against the weight of the edge. The values of α and β are adjusted such that the node having the edge with less weight and higher pheromone levels is selected.

After selecting the path, the ants will lay the pheromone according to Eq. 7.2.

$$\tau(i,j) = (1 - \rho) * \tau(i,j) + \rho * \tau_0 \qquad (7.2)$$

where ρ is the real valued coefficient such that $(1 - \rho)$ becomes the evaporation coefficient of the edge (i, j). The value of ρ must be between 0 and 1 i.e. $\rho \{\in\} (0, 1)$

The pheromone deposited by m ants is determined by:

$$\tau(i,j) = (1 - \rho) * \tau(i,j) + \rho * \Delta\tau(i,j) \qquad (7.3)$$

where

$$\Delta\tau(i,j) = \begin{cases} L_{gb} & \text{if } (i,j) \in \text{global best tour} \\ 0 & \text{otherwise} \end{cases}$$

The ant system has an important property of pheromone evaporation which causes the pheromone deposited to decrease over the period of time. This property helps in preventing premature convergence to a sub-optimal solution [9, 10]. Figure 7.1 shows the procedure followed while applying Ant Colony Optimization.

7.3.1 Ant Colony Optimization Applied to Task Scheduling in Cloud

To apply ACO for task scheduling in Cloud environment, it is necessary to plot the given scheduling problem in the form of a graph where nodes represent the operations and the execution time represents the edge weights [7]. Two dummy

7 An Equal Share Ant Colony Optimization Algorithm for Job Shop Scheduling... 85

Fig. 7.2 Plotting scheduling problem of nine tasks into graph

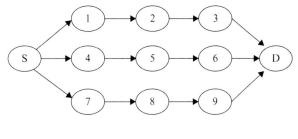

Fig. 7.3 Path used by Ant

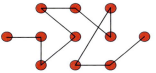

nodes, Start and Destination are added to the graph. Once the graph is plotted, the problem converts to Travelling Salesman Problem and we have to find the optimal path from Start to Destination.

In order to find the optimal path, ants are placed at the Start node, and are made to traverse the whole graph and reach the Destination node. While traversing, each ant is allowed to visit a node only once. Whenever the ant wants to move to next node, it calculates the probability of other nodes just like as in TSP where probability for each remaining node is calculated. The ant is allowed to choose the next node with highest probability. The ant uses formula (7.1) to calculate the probability of selecting the next path. After an ant chooses the node to visit, the selected node is marked as visited and the probability for remaining nodes is again calculated. The procedure is continued until all nodes are visited and the ant reaches the destination node. After the ant reaches the destination node, its path cost is calculated. The cost calculation is done by CloudSim. Figure 7.2 depicts an example path used by ant for moving from Start to Destination node while scheduling nine tasks and Fig. 7.3 shows an example path that is used by the ant.

7.4 Proposed Algorithms for Task Scheduling in Cloud Environments

The scheduling algorithm developed for JSSP is adapted to work for scheduling tasks in Cloud Environments. Two algorithms have been proposed for task scheduling in cloud environment. The first algorithm is hard to implement in practice, while the second algorithm gives better results and it is also practically implementable. The algorithms keep track of amount of execution carried out by each Virtual machine and makes decisions with a view to provide an equal share of tasks to all virtual machines. The performance of the algorithms tested using CloudSim.

```
begin
    Initialize all parameters
    Plot tasks into graph
    While (Stopping criterion not met)
        Place Ant on Initial node
        While (Stop when each ant has build a solution)
            Calculate Probability for each node
            Select Task with optimal probability
            Assign VM with least execution time(Load Balancing)
            Add execution time of current Task to selected VM's execution time
        end while
    end while
end
```

Fig. 7.4 Procedure for ACO for Cloud

Conventional ACO does not take into consideration the equal share factor, however the proposed Load Balancing ACO algorithm makes sure that each virtual machine gets equal amount of tasks to be scheduled. This helps in obtaining optimal solution. The two algorithms are:

(a) Task Length based Algorithm (TLA)
(b) Virtual Machine Attributes based Algorithm (VMAA)

7.4.1 Task Length Based Algorithm (TLA)

Figure 7.4 shows the procedure followed while applying Ant Colony Optimization to Task Scheduling in Cloud Environment.

While an ant is moving to the next node, the next node is selected based on the probability generated by the formula:

$$P(i,j) = \begin{cases} \dfrac{\tau(j)^\alpha * \eta(i,j)^\beta}{\sum_{u \in S_k} \tau(i)^\alpha * \eta(i,u)^\beta}, \end{cases} \tag{7.4}$$

where $\tau(j)$ represents the execution time of the task and $\eta(i, j)$ represents the pheromone trail corresponding to the path (i, j). Once the next task is selected, it is sent for execution on an ideal VM. The ideal VM is selected considering the load balancing factor. In this way, the whole graph is traversed. At the end, the execution time corresponding to current iteration is calculated and the pheromones are updated in inverse proportion to the execution time. This helps in making sure that the next iterations generate more accurate and optimal solution. The parameters α and β control the relative weight of task length and pheromone trails.

The above procedure is repeated for specified number of iterations after which an optimal solution is obtained. Thus, as we can see, ACO is based on constant improvement of solution until the best possible solution is obtained. Here the main target was

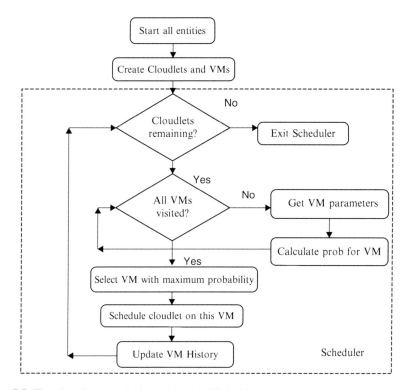

Fig. 7.5 Flowchart for task scheduling based on VM attributes

to design and implement an efficient ACO-based Task scheduling algorithm for Cloud Environment and test its results using CloudSim. The cost calculation i.e. the total time required for executing the given set of tasks is done using CloudSim.

7.4.2 Virtual Machine Attributes Based Algorithm (VMAA)

The procedure for algorithm based on VM attributes is expressed in the Fig. 7.5. First of all, entities of CloudSim are started. Tasks are named as Cloudlets in CloudSim. Cloudlets are created and also virtual machines are created. Then, control is passed to Scheduler module which is shown in dotted rectangle. This scheduler is responsible for assigning appropriate virtual machine to a specified task.

When a new task enters the scheduler, the probability module starts playing its role. Here, the probability for each virtual machine is calculated. For calculating the probability, it is necessary to get the corresponding parameters of the virtual machine. After calculating probabilities of all virtual machines, the virtual machine with maximum probability is selected. This virtual machine is then assigned the job of executing the task. The same procedure is repeated for each incoming cloudlet (task).

If there are no more cloudlets remaining, the control exits the scheduler and the task scheduling is completed.

The rule for calculating the probability determines the effectiveness of the proposed algorithm. The formula incorporates factors like priority to high configuration virtual machines, prevention of starving of low configuration machines and load balancing.

$$P(i) = \begin{cases} \dfrac{\tau(i)^{\alpha} * \eta(i)^{\beta}}{\sum_{u \in S_k} \tau(i,u)^{\alpha} * \eta(i,u)^{\beta}}, & \text{if } j \in S_k \\ 0, & \text{otherwise} \end{cases} \qquad (7.5)$$

where, $\tau(i)$ indicates the processing capacity of virtual machine. This processing capacity is calculated based on:

$$P = num\_proc * ram * mips \qquad (7.6)$$

Where *num_proc* indicates the number of CPUs present in the virtual machine, *ram* indicates the amount of physical memory present in virtual machine and *mips* is the MIPS of the virtual machine.

The parameter $\eta(i)$ helps in load balancing of virtual machines. A priority is assigned to each virtual machine which is reduced after a machine is assigned any task. Due to this priority, the machines with low configuration get a good probability of getting a task. This indeed helps in preventing any sort of starvation of virtual machines for tasks. Also, to achieve effective distribution of tasks over virtual machines, a variable *index* is used and virtual machines are given priority if their id equals index. The variable index keeps on occupying values from 0 to size of virtual machine list.

7.5 Results and Discussion

7.5.1 Results for Task Length Based Algorithm (TLA)

The experiment was carried out using CloudSim – a cloud simulator. The task size was kept around 10K instructions for smaller sizes and it was increased to one million instructions for considering heavy tasks. In these experiments, the values of parameters are very important. With the experiments carried out, it has been seen that $\alpha = 0.1$, $\beta = -2.0$, $\rho = 0.01$ gave optimal values. Also, the load balancing function helped in achieving optimal solutions. The experiment gave good results after 100 iterations in most cases. So, each reading was taken after 100 iterations. Table 7.1 presents a comparison of readings obtained for First Come First Served (FCFS) algorithm, basic ACO algorithm and our proposed Load Balancing algorithm.

7 An Equal Share Ant Colony Optimization Algorithm for Job Shop Scheduling...

Table 7.1 Comparison of makespan of Load Balancing ACO with conventional ACO and FCFS for cloud task scheduling

		Observed makespan times (best case)		
n (no. of jobs)	m (no. of machines)	FCFS	Traditional ACO	Load Balancing ACO
9	3	253.3	242.5	241.4
40	4	324.1	284.89	272.8
81	9	1,848.2	1,748.7	1,289.1
400	10	1,842.2	1,645.9	1,136.1

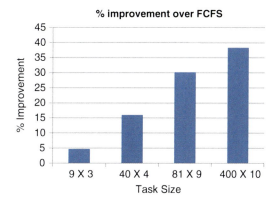

Fig. 7.6 Comparison of Load Balancing ACO over FCFS

As seen from the Table 7.1, for smaller problem sizes with less number of tasks and machines, traditional ACO outperformed FCFS. But, load balancing ACO got the results just near traditional ACO. Still, the readings obtained were better than traditional ACO. But, for complex problems, the difference in reading of traditional ACO and Load Balancing ACO increased significantly and Load Balancing ACO clearly outperformed other methods by a large margin. As we can see, for 400 tasks problem, the load balancing ACO obtained optimal solution of 1,136.1 time units while traditional ACO got 1,645.9 as output.

Table 7.1 and Fig. 7.6 show the percentage improvement obtained by our proposed algorithm based on ACO over FCFS. It is clear that as the problem complexity increases i.e. as the number of tasks and virtual machines increases, the performance improvement increases. Here, the problems are represented in the form: n × m. n × m indicates n jobs to be scheduled on m machines. Thus, we can see from above chart that Load Balancing ACO clearly outperforms FCFS for tasks of all sizes and complexities. The X-axis indicates the problem size. Load balancing ACO performs 4.7 % better than FCFS for 9 × 3 problem. It performs 16 % better for 40 × 4, 30.2 % better for 81 × 9 and 38.32 % better for problem of size 400 × 10. Thus, it can be easily concluded that the proposed algorithm outperforms FCFS and even basic ACO for task scheduling.

Thus, experimental results showed that our proposed algorithm outperforms conventional ACO. This performance improvement increases as the complexity of the problem grows. Thus, addition of load balancing factor helped in assigning equal load to machines and the effective makespan time decreased.

Table 7.2 Comparison of FCFS and Load Balancing ACO for cloud task scheduling

		Observed makespan times (best case)	
n (no. of jobs)	m (no. of machines)	FCFS	Load Balancing ACO
5	3	186.1	145.3
10	4	258.2	188.2
20	5	346.2	258.1
140	5	2,413.9	1,745.3
400	7	2,423	1,564.7

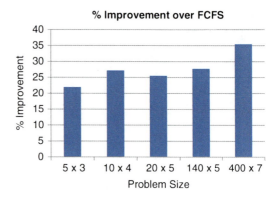

Fig. 7.7 Comparison of Load Balancing ACO over FCFS

7.5.2 Results of VM Attributes Based Algorithm (VMAA)

The experiment was carried out using CloudSim – a cloud simulator. The task size was kept around 10K instructions for smaller sizes and it was increased to one million instructions for considering heavy tasks. In these experiments, the values of parameters played an important part. With the experiments carried out, it has been seen that $\alpha = 0.1$ and $\beta = 4.0$ gave optimal values. Also, the load balancing function helped in achieving optimal solutions. Table 7.2 presents a brief comparison of readings obtained for First Come First Served (FCFS) algorithm and the proposed Load Balancing algorithm.

As it can be seen from Table 7.2, the proposed Load balancing ACO outperformed FCFS in each and every reading, with each task size complexity. The first four readings were recorded for light tasks. As we can see, for 5×3 problem size, FCFS required 186.1 time units while the proposed algorithm completed the execution in just 145.3 time units. Similarly, for 10×4 problem, FCFS recorded makespan of 258.2 while Load Balancing ACO showed result of 188.2 time units. The same happened for 140×5 problem where Load Balancing ACO outperformed with 1,745.3 time units against 2,413.9 time units of FCFS. The last reading of 400×7 problem size was taken with heavy tasks of size one million instructions to test the algorithm for heavy tasks. The percentage improvement of proposed algorithm over FCFS is shown in the Fig. 7.6.

Figure 7.7 shows the percentage improvement obtained by our proposed algorithm based on ACO over FCFS. It is clear that as the problem complexity increases i.e. as the number of tasks and virtual machines increases, the performance improvement increases.

Here, the problems are represented in the form: n × m. n × m indicates n jobs to be scheduled on m machines. For 5 × 3 problem size, the Load Balancing ACO showed an improvement of 21.92 % over FCFS, while for the 10 × 4 problem size, the improvement was 27.11 %. The proposed algorithm reduced makespan by 25.4 % for problem of size 20 × 5 while the performance increased to 27.7 % for problem of size 140 × 5. Thus, it is evident from this case that if the number of tasks is increased, the performance also increases. The final problem 400 × 7 showed improvement of 35.42 % over FCFS, making it clear that the proposed algorithm based on Load Balancing ACO outperformed FCFS in almost every problem size.

7.6 Conclusion

In this paper, we have proposed two Load Balancing ACO algorithms for achieving better results in cloud task scheduling problem. We have also evaluated and compared the results obtained with conventional ACO algorithm and FCFS and found that our proposed algorithms give better performance. As the size of scheduling problem increases, the difference in the results produced by both these algorithms increases and Load Balancing ACO algorithms proved to be more useful. The algorithm based on Task Length is particularly useful for finding the best solution for a given scheduling problem. The algorithms have shown better performance in all the tested cases.

References

1. A. Jain, S. Meeran, Deterministic job-shop scheduling: past, present and future. Eur. J. Oper. Res. **113**, 390–434 (1999)
2. D. Merkle, M. Middendorf, H. Schmeck, Ant Colony Optimization for resource constrained project scheduling. IEEE Trans. Evol. Comput. **6**(4), 333–346 (2002)
3. C.A. Silva, J.M. Sousa, T. Runkler, JMG. Sá da Costa, A logistic process scheduling problem: genetic algorithms or ant colony optimization, in *Proceedings of 16th World Congress of the International Federation of Automatic Control, IFAC*, Czech Replublic, 2005, pp. 1–6
4. M. Dorigo, T. Stutzle, *Ant Colony Optimization* (The MIT Press, Cambridge, MA, 1992)
5. M. Dorigo, V. Maniezzo, A. Colorni, Ant system: optimization by a colony of cooperating agents. IEEE Trans. Syst. Man Cybern. B Cybern. **26**, 29–41 (1996)
6. M. Den Besten, T. Stützle, M. Dorigo, Ant colony optimization for the total weighted tardiness problem, in *Parallel Problem Solving from Nature PPSN VI*, Springer, Berlin/Heidelberg, 2000, pp. 611–620
7. Hui Yan, Xue-Qin Shen, Xing Li, Ming-Hui Wu, An improved Ant Algorithm for job scheduling in grid computing, in *Proceedings of the Fourth International Conference on Machine Learning and Cybernetics*, Guangzhou, 18–21 Aug 2005

8. K.R. Ku-Mahamud, H.J.A. Nasir, Ant Colony Algorithm for job scheduling in grid computing, in *Fourth Asia International Conference on Mathematical/Analytical Modelling and Computer Simulation*, Kota Kinabalu, Malaysia, 2010, pp 40–45
9. Jun Zhang, Xiaomin Hu, X. Tan, J.H. Zhong, Q. Huang, Implementation of an Ant Colony Optimization technique for job shop scheduling problem. Trans. Inst. Meas. Control, **28**, 93–108 (2006)
10. M. Dorigo, L.M. Gambardella, Ant Colony System: a cooperative learning approach to the travelling salesman problem. IEEE Trans. Evol. Comput. **1**, 53–66 (1997)

Chapter 8
A Dynamic Trust Computation Model for Peer to Peer Network

Tarek Helmy

Abstract Reliable Peer to Peer (P2P) communication has always been a challenging task for P2P system designers. Trust models try to alleviate problems in P2P systems by incorporating a variety of approaches and schemes. However, most of the current models measure the trustworthiness of a peer only by its trust value which results in inefficient mechanisms of dealing with malicious peers. This paper proposes a unique way of computing the trust value of peers in two steps; by computing the trust value of a peer after each transaction and computing the trust value after a periodic interval of time, namely the transactional trust and revised trust respectively. In the proposed P2P architecture, peers are distributed into groups and each group has a central peer which is responsible for the peers in its group. A management peer is used to manage the central peers and it takes care of all other management activities in the system. The simulation results validate the fact that the proposed trust computation model is accurate and computes the trust efficiently.

8.1 Introduction

P2P systems have become a very popular area of research in the recent decade. These systems offer numerous services which include file sharing and storing, distributed computing, collaborative applications and immediate communication. Nevertheless, existing systems are susceptible to various security problems for the

---

Tarek Helmy on leave from Faculty of Engineering, Department of Computers & Automatic Control Engineering, Tanta University, Egypt

T. Helmy (✉)
Department of Information and Computer Science, College of Computer Science and Engineering, King Fahd University of Petroleum and Minerals, Mail Box 413, Dhahran 31261, Kingdom of Saudi Arabia
e-mail: helmy@kfupm.edu.sa

scale of the network and unreliable nature of peers characterizing most P2P systems. Among the heterogeneous peers, some might be honest and provide high-quality services, some might be self-serving (for example free-riders) and don't want to provide service to others, some might be even malicious ones who provide bad service and harm the customers. Historical behavior of the peer is used to build trust value for a peer. A peer can estimate the authenticity of files other peer share by their trust value. Generally, the bigger the trust value, the higher is the probability that the files shared are authentic. But the idea of using trust value as a parameter to judge has become infamous as some peers act as traitors, where a peer develops a high trust value by providing good service and then suddenly starts giving bad service for some transaction. Calculation of trust also arise many parameters to be taken care of, as the peers may collude with each other to provide a good rating to bad peer and a bad rating to good peer. So the idea of using only peer's trust value has become stale. So it makes necessity to include many other parameters (like vote accuracy factor) for accurate calculation of trust. In the proposed model, we use the group based approach where the peers are divided into groups. Each group has a fixed number of peers and a Central Peer (CP) which is responsible for calculating the trust values for the peers in its group. We also introduce a new parameter "Likelihood of Defecting" (LD) which gives the probability of the peer defecting on a transaction (it is the average number of unsuccessful transactions of a peer). The higher the value of LD, the greater is the chance of transaction failure or getting a bad service from a peer. We use a dynamic two-step model for calculating the trust for a peer. In the first step, transactional trust is calculated after every transaction. It is incremented or decremented based upon the success or failure of the transaction. In the second step, the revised trust is calculated after periodic intervals of time. The rest of the paper is organized as follows. Section 8.2 presents taxonomy of few existing trust computation models. Section 8.3 presents the architecture of the proposed trust computation model. Section 8.4 describes the transactional forms and tables used in the computation of the trust values. Section 8.5 explains the actual calculation of transactional trust, revised trust and LD values. Section 8.6 presents the results. Section 8.7 concludes the paper and highlights the future work.

8.2 Taxonomy of Existing Trust Models

Due to diverse characteristic of peers, some peers might be malicious which effect the development of P2P network. Various models have been proposed to cope up with this problem. Wang et al. [1] proposed a new trust model for e-commerce security based on voting agreement. It prevents vulnerable peers from involving in transaction and the problem of no-history record peers is also addressed. The peers are evaluated based on two-side trading experience and recommendation of other peers. Xu et al. [2] proposed an enhanced trust model based on reputation. Trustworthiness of peers is evaluated based on direct interaction experience and other peer's recommendation. Voting for peers is considered from trust and distrust perspective. The model solves the trust problem of a new peer and is effective to prevent the malicious attacks. Ning Liu et al. [3] facilitate authentication and trust between peers in P2P systems along with enhanced security of the model. Cuihua Zuo et al. [4] proposed a multi-level trust model which controls the peer's access permission by their level values. The simulation results justify that the model is effective to prevent malicious act and encourage peers to share their resources in P2P network. Hongcai Feng et al. [5] proposed a new security policy based on bi-evaluation of trust and risk. The access permission of transactions among peers in P2P file-sharing is controlled by applying the trust and risk factors. All the previous approaches consider peers to give trust values for other peers in the system. They consider factors like voting factor, reliability, risk factor and other parameters for accurate calculation of trust but none of them have used the concept of Network Monitor (NM) in trust systems and also additional factors like the likelihood of defection.

8.3 System Architecture

Peers constitute the major part of any P2P system. The proposed system architecture is group based, which implies that peers are distributed across the system in groups. Each group has a CP which is responsible for all the peers in the group. The number of peers which can be accommodated in a group is set by the initial results obtained by considering the workload and overhead of managing these peers by the CP. As stated earlier, apart from the groups and the CPs in the P2P reputation system, there is a Management Peer (MP), a Network Monitor (NM), Network Statistic Store (NSS), an Authentication System (AS), a Trust Calculation (TC) system and a Global History (GH) reputation system. We will discuss all these components in the following sub-sections. Figure 8.1 shows the complete system architecture.

8.3.1 Central Peers

These peers are preset by the P2P system designers. They are not involved in any transactions and exchange of information with the other peers in the system. Their job is to manage all the other peers in their groups.

The most important job of a CP is the calculation of trust values of all the peers in its group. This is a relatively difficult task as each CP has to follow the conventions of TC which are defined by the TC system. It involves maintaining trust tables, populating the required fields and then computing the trust values using a certain formulae. What are the fields of the trust tables, how many trust tables are required and their information will be discussed in Sect. 8.4. Another

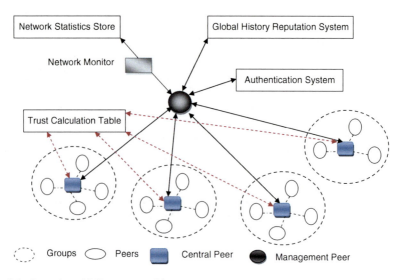

Fig. 8.1 Group based P2P system architecture

job of the CP is to allow smooth interaction and transactions between peers of different groups. The CP synchronizes with the MP and helps establishing the link between peers of different groups. It is intuitive that all CPs are connected to the MP. Additionally, the CP updates all the trust values to the MP which validates the correctness of the calculated trust. Lastly, the CPs act as sources of information for the newly joined peers or for any peer requesting the trust value of another peer in the group before transacting with them. There is no election for the CPs of a group as they are preset by the system. There are no malicious and trust concerns about the CPs as they are not involved in any transactions. Their sole purpose is the management of the group.

8.3.2 Management Peers

The MP is the central entity of the P2P system which is responsible for binding different groups together. There are various responsibilities associated with a MP. First, the MP is responsible for any new peer entering the system. It assigns each peer a unique system identifier which is the peer ID. The peer ID cannot be changed throughout the life time of a peer in that system. The peer ID is coupled with the peer's login name, password and verification details such as credit card number. An authentication table is maintained which keeps information about all the peers in the system. A login server authenticates the entry of peers in the network. This is done to minimize the possibility of white washing. All this constitutes the AS which is accessible to the MP. Second, the MP facilitates communication between peers of different groups by synchronizing with the CPs of the groups. Third, the MP contributes in the trust calculation by the CPs in a way that, when a CP calculates the trust of a peer based on the transaction forms filled by the transacting peers, in case of a mismatch between the information in the forms, the CP refers the case to the MP which in turn checks the NSS for the particulars of that transaction. It then decides which peer was truthful and notifies the CP to calculate the trust accordingly. The details of the transaction forms will be explained separately in the following sections. Finally, whenever a CP updates the calculated trust to the MP, it transfers the data to the GH system as it is the only entity in the system with access to GH system. Similar to the CPs, there is no election for the MP and it doesn't take part in any transactions. It is a preset entity nominated by the system. Alternatively, in case of a failure there is a backup MP which offsets the failure of the MP.

8.3.3 Network Monitor and Network Statistics Store

The use of a NM has been incorporated in the proposed approach for verification and validation purposes. The NM captures all the network traffic between peers according to the transaction identifiers. All the information is stored in the NSS. As mentioned earlier in the responsibilities of MPs, this stored information is much helpful in verifying the validity of data filled in the transaction forms by the transacting peers. On referral by a CP, the MP verifies the transaction form with the information available about that transaction in the NSS. Basically the attempt is to capture the number of packets exchanged between two peers or in other words, the size of the data traffic in that transaction. This helps in analyzing which peer correctly filled the transaction form. Based on this, the CP calculates trust of a peer by rewarding the truthful peer by incrementing its trust value and punishing the malicious peer by decrementing its trust value. The NM services can be customized and used according to the requirements of the system under consideration. The use of this component is aimed at attempting to distinguish between good and malicious peers.

8.3.4 Global History System

Among various sources of information gathering in a P2P reputation system, the most comprehensive solution is the use of a GH reputation system. Other sources of information can be friends of real world, friends on social networking sites, transitive trust chains, peers with which successful transactions were realized and so on. But among all, the GH reputation systems are most reliable. In the cases when the probability of a single fraudulent opinion is greater, the collective sum of all opinions is comparatively more accurate even when a large fraction of peers are malicious [6]. Generally, the quality and quantity of information are diametrically opposite. In our case, the GH reputation is developed to provide comprehensive solutions for peers requesting information about other peers in the network. The computed trust values are transferred by the CPs to the MP who in turn populates the data tables of the GH system. One of the most comprehensive reputation systems is that of eBay [7], which consists of a single trusted entity that collects all transaction reports and rates all users.

8.4 Transaction Forms and Tables

In this section, various tables used by different components will be discussed in depth. Other algorithms like [8–10] use the concept of weighted transaction reports to gather information about peers before computing the trust values for peers. A transaction is defined as the interaction between peers. The interactions can include swapping files, storing data, answering queries or trading items involving monetary payments [6]. The peers involved in a transaction are called transacting peers. After every transaction both the source peer and the destination peer are required to fill a transaction form providing feedback about the details of the transaction. The source peer is the peer who initiates or requests the file and the destination peer is the one which responds or shares the file. Both peers fill the transaction form and submit it to the CP of the group. The CP uses the data from the transaction forms for computing the trust value of both peers. It is to be recalled that in case of a mismatch between the transaction forms of the source and destination peers, the NSS is referred to verify the correctness of either of the peers and consequently the trust is calculated.

8.4.1 Transaction Forms

A transaction form consists of various attributes. These are filled by the source and destination peers involved in a transaction. There are a total of ten attributes in a transaction form out of which five are prefilled by the CP before giving the form to transacting peers. The remaining five attributes are filled by the transacting peers. The attributes are as follows. **Transaction ID** (T_{id}) is the transaction identifier which is unique for every transaction. It is allotted by the CP for every transaction. It is important because it is the primary key for various tuples in various tables used by multiple components. **Source Peer ID** is a unique system identifier associated with the source peer. Each peer has a unique system identifier allocated to it by the MP upon entry into the system. This attribute is prefilled by the CP in the transaction form. **Destination Peer ID** is a unique system identifier associate with the destination peer. This attribute is also prefilled by the CP in the transaction form. These attributes are prefilled by the CP because it may happen that malicious peers may fill a random Peer ID's in the forms which may cause problems in scoring and ranking the peers. **Start Time Stamp** is the start time of the transaction. It is

Table 8.1 A possible assignment of the Transaction Importance values

File size	Monetary value	T_i
1–10 MB	1–50$	0.2
10–100 MB	50–500$	0.4
100–1,000 MB	500–1,000$	0.7
>1 GB	>1,000$	1.0

basically used to match the information about a transaction with the information retrieved from the NSS. Also, it is of high importance in calculating the time factor which is used in the computation of revised trust values. This attribute is prefilled by the CP in the transaction form. **End Time Stamp** is the end time of the transaction. Its significance is the same as that of the start time stamp. It is pre-filled by the CP in the transaction form. **Transaction Completeness** ($T_{c/l}$) attribute helps the source peer to report any inconsistency in the transaction. The transaction completeness is the degree of satisfaction the requesting peer experienced in a particular transaction. Its value is between -1 and $+1$, where -1 represents dissatisfaction or incompleteness of transaction and $+1$ represents satisfaction or completeness of transaction. **Transaction Importance** (T_i) attribute represents the importance of a particular transaction. It is helpful in calculating the transactional trust of a peer. It contributes proportionally to the transactional trust of a peer. Its values are pre-defined by the system. Table 8.1 shows a possible assignment of transactional importance values. These values are customizable and can be set according to the system under consideration.

Transaction Count (T_c) represents the number of interactions between a set of peers. If a specific peer interacts with the same peer 'n' times, then the transaction count is set to 'n'. This is used to take into account the transactional experience of a peer with another peer. It is used in the revised trust computation. The transaction count is an integer value.

Transaction Type attribute specifies the type of transaction. The type of a transaction can be file sharing, monetary transaction, or answering queries. **Transaction Size** is the size in bytes of the file shared between peers. This is recorded because in case of a mismatch between transaction forms of transacting peers, the size of transaction is retrieved from the NSS and compared with the transaction size attribute of both the source and destination transaction forms. This helps in catching the malicious peers and punishing them accordingly.

8.4.2 Tables Involved in the Computation Process

Authentication Tables are maintained by the AS. They are accessible to the MP and store information about all the peers in the P2P system. The main purpose is to disallow malicious peers from disconnecting themselves from the system and then rejoining the system at a later stage. Also, it helps peers from using multiple fake Peer IDs. This alleviates the problem of white washers and Sybil attacks. Table 8.2

Table 8.2 Authentication table

Peer ID	User name	Password	Credit card number
p2pnum1	abc	xxxx	9824 3561 xxxx xxxx
p2pnum2	xyz	xxxx	9745 6666 xxxx xxxx

Table 8.3 Network statistics table

Transaction ID	Source peer ID	Destination peer ID	Start time stamp	End time stamp	Data size (B)
54892	p2pn1	p2pn19	04.00 A.M.	04.12 A.M.	25.5 MB

Table 8.4 Transactional trust table

T_{id}	S_{id}	D_{id}	T_c	$T_{c/l}$	T_i	T_a	TR

Table 8.5 Revised trust table

Peer ID	T_{id}	TR	T_c	Z	T_{rev}

Table 8.6 Abstract trust table

Peer ID	Trust value	Likelihood of Defecting – L(d)

shows the various fields of an authentication table. The use of login servers helps prevent the system from unauthorized and unauthenticated access. This limits and puts a check on the adversarial powers of peers.

Network Statistics Tables are maintained by the NM component. They consist of fields which are essential in capturing the required information pertaining to a transaction. These tables are very helpful in comparing the data of transaction forms and verifying validating their correctness. Table 8.3 shows the various fields of the network statistics tables.

Trust Tables are the most essential tables maintained by the CPs. The central peers fill the fields of these tables using the transaction forms. The trust values are calculated using appropriate formulae. There are three tables associated with trust computation. Since, the computation of trust is a two-step process, the first table (Table 8.4) accounts for the calculation of the Transactional Trust, whereas the second table (Table 8.5) accounts for the calculation of Revised Trust and the third table (Table 8.6) contains the abstract information of trust values of peers. The data of the first two tables are transferred to the MP in order to be populated in the Global Reputation System. The third table is maintained by the CP itself, so that when a peer in its group requests for the trust value of a peer, it can send the corresponding trust value. Another factor called LD is included in the abstract trust table, which will be discussed in the following section. **Global History Reputation System Tables** contains the most comprehensive information about a peer in a P2P reputation system. The tables maintained by the global system are simply the union of all fields in various tables across different system components. The table can be represented as follows: GH System Table = (Authentication Table) ∪ (Network Statistics Table) ∪ (Transactional Trust Table) ∪ (Revised Trust Table) ∪ (Abstract Trust Table).

8.5 Trust Computation

Once all the tables are populated with the corresponding information, a reputation score or rather the trust value of a peer is computed. Many authors have proposed various methods to compute the trust of a given peer. The computation of trust depends on many factors like satisfaction, experience, and transactional importance [1]. Few models reduce these factors to one or two like trustworthiness of a peer and its contribution [5]. Some of the optimal factors to be considered for calculating a peer's trust value are specified in [6]. The choice of factor depends on the relevance of the factor with reference to the method of trust computation. We propose a unique two-step trust computation model which uses the most essential factors required in calculating a peer's trust value. As specified earlier, the trust computation is done by the CPs of each group.

8.5.1 Transactional Trust

The CP takes all the necessary information from the trust tables and computes the trust value of each peer in its group. After every transaction in the system, the CP calculates the trust of the source and destination peer as per the information in the transaction forms and tables. We discuss two cases of incremental and decremental transactional trust. An important point worth mentioning is that the range of the final trust value of each peer is between 0 and 1. The increments and decrements are a fractional value corresponding to first and second decimal places as we will see in the following calculations. Initially, at the start of the computation, each new peer joining the system is assigned a value of 0.5 which increments or decrements corresponding to its good or malicious behavior. **Case 1:** The source peer can be malicious or non malicious. **Case 1a:** If the source peer was not malicious and it filled the transaction form correctly, there will be no change to its trust value. The trust value would remain same as the previous transactional trust value. **Case 1b**: If the source peer is found to be malicious. This is the case, if there was a mismatch in the transaction forms and the source peer was found to be malicious after comparing with the NSS information. The decremental value is computed proportionally to the importance of the transaction which was specified in the transaction form. Defecting on a higher importance transaction will decrement the peer's trust value more and subsequently defecting on a lesser importance transaction will decrement the peer's trust value less. For simplicity purpose, we fix the decremental value to be one tenth of the transactional importance, i.e. if the transactional importance in a transaction was 0.7 and the source peer defected on it, then its trust value is decremented by 0.07. **Case 2:** The destination peer can be malicious or non-malicious. **Case 2a:** If the destination peer is found to be non-malicious. Then, we have to increment the peer's trust value using the following formula: $TR_i = TR_{i-1} + T_a$, where, TR_i is the transactional trust of the peer with respect to the current transaction, TR_{i-1} is the trust value of the peer before that transaction. Initially for a new peer in the system, $TR_{i-1} = T\ initial = 0.5$. The actual trust for the current transaction is denoted by T_a and calculated as: $T_a = (TC/I * T_i)$ where TC/I implies the satisfiability of a source peer with the destination peer's service and T_i is the importance of that transaction. It is notable that the product of these two factors is a value corresponding to second decimal place. This notifies the important observation that, the transactional trust is incrementally or decrementally updated after each and every transaction of the peer in the system. This also highlights the non-consideration of other trust related factors such as time and experience of a peer in calculating the trust value. For the same reason, we propose the second step in the computation of a peer's trust value, namely the Revised Trust. **Case 2b:** Alternatively, the destination peer can be malicious. This is the case, if there is a mismatch in the transaction forms and the destination peer or the responding peer is found to be malicious. We follow the same methodology as used in Case 1b for malicious source peers. The T_a value is calculated as one tenth of the importance of the transaction and deducted from the previous transactional trust value.

8.5.2 Revised Trust

Revised trust is calculated for all peers in the system after a periodic interval of time. This considers the time of the transactions and the experience of each peer in the system. The revised trust is calculated using $T_{rev} = \Sigma i = 1\ to\ n\ (TR_i * Z_f)/T_c$, where, T_{rev} is the revised trust value of a peer calculated by averaging the peer's transactional trust values in individual transactions over a period of time. Z_f is the final time factor which is used to include only the most recent transactions. The revised trust value is averaged over the number of transactions considered and hence transaction count T_c is used. The initial time factor 'Z_i' is calculated using $Z_i = (t\ current - t\ transaction)$, where, t current and t transaction are the current time and time of the particular transaction respectively. The lesser the time difference implies recency of the transaction and larger the time difference implies oldness of the transaction. Since, recent transactions should be given more weight; we map the Z_i to Z_f values by assigning a weight of 1 to seven most recent transactions, a weight of 0.9 to next seven recent transactions, a weight of 0.8 to next seven recent transactions and so on. The revised trust value is updated by the CP after a periodic interval of time which can be set by the system. The CP stores this value in the Abstract Trust Table (ATT) and keeps updating it periodically. Note that, the transactional trust value is not updated in the ATT. It is to be recalled that the abstract trust table had another field namely LD.

8.5.3 Likelihood of Defecting

It has been mentioned by many authors that only the trust value is not enough for determining a peer's reputation in the system. We propose the LD factor takes into account the probability that a peer will defect on a certain transaction irrespective of its trust value. The LD is calculated as follows: L(d) = (Number of failed transactions in a set of recent transactions)/(Total number of recent transactions). The number of recent transactions to be considered for calculating L(d) is decided by the system. The L(d) value is stored in the ATT with the CPs. Whenever a peer in the group or system rather, requests the CPs for the trust value of another peer it wishes to transact with, the CP sends the two field tuple of 'trust value' and 'L(d)' from the ATT to the peer. An important observation is that, the use of the LD factor solves the problems of traitors in the system. With the use of L(d), this will not be the case, as the number of recent defects or transactional failures will be accounted. If a peer suddenly goes defecting on transactions, the recent transactions can be considered for calculating L(d) and correspondingly it will give a higher value. The higher value of L(d) implies the traitor behavior of a peer in the system.

8.6 Experimental Results

For the purpose of analyzing the proposed trust model, we used Matlab7.0 to simulate the P2P environment. The simulation results validate the performance of the proposed computational model. The simulation demonstrates the transaction success and the calculation of trust values based on the proposed parameters for all the peers in the system. For the purpose of simulation, we have used a single group as the sole entity in the P2P system. The same computational model can also be applied to all groups in the system. There are 100 peers in the group and each peer transacts with exactly 10 other unique peers. Among the group there are x % malicious peers who provide un-trusty ratings for the other peers with whom they interact. The peers transact with each other and after each transaction, their trust values are updated continuously. Initially each peer starts with a trust value of 0.5. As the transactions increase, the trust values are updated more frequently. The 'revised trust' parameter takes care of the periodic calculation of trust value for each peer. The number of transactions in the simulated system is 1,000. This is repeated for 50 iterations which makes the total number of transactions on which results are analyzed to 50,000. Figure 8.2 shows the averaged trust value of peers in the system over 50 iterations by including different values of un-trusty peers within the system. We have used 4 different values of un-trusty settings to analyze the differences in the computed trust values. The results validate the fact that the trust computation model is accurate and computes the trust efficiently. When there are only 10 % un-trusty peers in the system, the average trust values of all the 100 peers in the system is high, i.e. between 0.9 and close to 1. Similarly, when there are 30 %

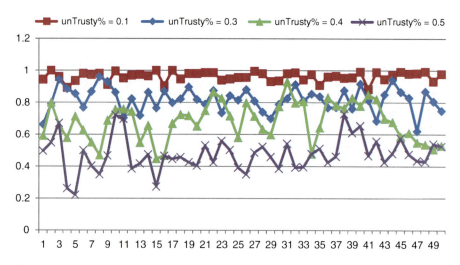

Fig. 8.2 Trust values of peers for various settings of number of un-trusty peers in the system

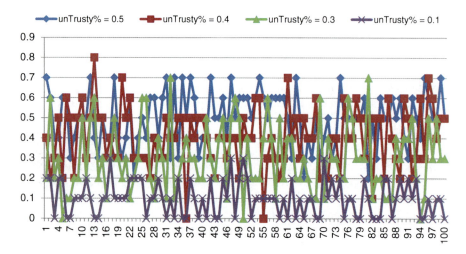

Fig. 8.3 Likelihood of Defecting values for 100 peers in the system

un-trusty peers in the system, the average trust value of the peers is in the range of 0.7 and 0.9 with the lowest being 0.625 and highest being 0.95. Analyzing the same for the setting where the number of un-trusty peers are 50 % in the system, the average trust value of the peers are in the range 0.4 and 0.7 with the lowest being 0.225 and the highest being 0.75.

Figure 8.3 shows the plot of the newly proposed parameter 'LD' with respect to all the 100 peers in the system for 4 different settings of the number of un-trusty peers (x %) in the system. When the number of un-trusty peers in the system is

10 %, the LD is in the range of 0–0.2. When the number of un-trusty peers is increased to 30 %, the LD is in the range of 0.2–0.4. Similarly, when the number of peers in the system is 50 %, the LD is in the range of 0.4–0.7. This clearly explains that as the number of un-trusty peers increase in the system, the LD also increases proportionally. As stated earlier, it is a valuable parameter in discarding the trust ratings received from such peers in the system. The current work is well structured and the experimental results validate the outcomes of the analysis.

8.7 Conclusion and Future Work

This paper presents a dynamic model of trust computation in P2P systems using a centralized approach. The structure of the P2P system is group based. The model uses a network monitor to monitor the transactional activity of the system and contributes to the development of a global reputation system. Various adversaries such as white washers and traitors are taken care by the use of the AS and the LD factor. The trust values are updated accurately after each transaction and revised on a periodic basis. The model is simple, efficient and reliable in terms of a P2P network environment. Our future direction will be towards implementing both the network statistics store and the global reputation system then scaling up the number of peers and groups and measure the efficiency of the proposed model.

Acknowledgment This work is supported by King Abdulaziz City for Science and Technology through King Fahd University of Petroleum & Minerals Science & Technology Unit under the project No. 10-INF1381-04 as part of the National Science, Technology and Innovation Plan. Thanks extended to Mr. Mohsen and Mr. Wajhat for their help in running and monitoring the simulation model.

References

1. Wang Yu, Zhao Yue-long, Hou Fang, A new security trust model for peer-to-peer E-commerce, in *Proceedings of IEEE International Conference on Computer Science and Software Engineering*, 2008, pp. 797–801
2. Xu Wu, Jingsha He, Fei Xu, An enhanced trust model based on reputation for P2P networks, in *Proceedings of IEEE International Conference on Sensor Networks, Ubiquitous and Trustworthy Computing*, 2008, pp. 67–73
3. Ning Liu, Jianhua Li, Liming Hao, Yue Wu, Ping Yi, Group-based trust model in P2P system based on trusted computing, in *Proceedings of IEEE International Conference on Computer Science and Software Engineering*, 2008, pp. 797–801
4. Cuihua Zuo, Hongcai Feng, Jianfang Zhou, A security policy based on bi-evaluations of trust and risk in P2P systems, in *Proceedings of 2nd IEEE International Conference on Education Technology and Computer*, Vol. 5, 2010, pp. 304–309
5. Cuihua Zuo, Hongcai Feng, Jianfang Zhou, A novel multi-level trust model to improve the security of P2P networks, in *Proceedings of 3rd IEEE International Conference on Computer Science and Information Technology*, 2010, pp. 100–104

6. S. Marti, H. Garcia-Molina, Taxonomy of trust: categorizing P2P reputation systems. Comput. Netw. **50**(4), 472–484 (2006)
7. eBay: The Worlds Online Marketplace. Available from: http://www.ebay.com/
8. S.D. Kamvar, M.T. Schlosser, H. Garcia-Molina, The EigenTrust algorithm for reputation management in P2P networks, in *ACM Proceedings of the Twelfth International World Wide Web Conference*, 2003, pp. 640–651
9. L. Page, S. Brin, R. Motwani, T. Winograd. The PageRank citation ranking: bringing order to the web, Technical Report, Stanford Digital Library Technologies Project, 1998
10. A. Das, M.M. Islam, SecuredTrust: a dynamic trust computation model for secured communication in multiagent systems. IEEE Trans. Depend. Secure Comput. **9**(2), 261–274 (2012)

Chapter 9
Road Traffic Management Using Egyptian Vulture Optimization Algorithm: A New Graph Agent-Based Optimization Meta-Heuristic Algorithm

Chiranjib Sur and Anupam Shukla

Abstract In this paper we have continued the introduction and application of a new nature inspired meta-heuristics algorithm called Egyptian Vulture Optimization Algorithm (EVOA) which primarily favors combinatorial optimization problems and graph based problems. The algorithm is derived from the nature, behavior and key skills of the Egyptian Vultures for acquiring food for leading their livelihood. These spectacular, innovative and adaptive acts make Egyptian Vultures as one of the most intelligent of its kind among birds. The details of the bird's habit and the mathematical modeling steps of the algorithm are illustrated demonstrating how the meta-heuristics can be applied on the route planning for a graph based road network depending on the multi-parametric optimization of distance (travel time) and waiting time. Due to the dynamically changing behavior of the waiting time for the various crossings of the network, the system is dynamic system and the best optimized path tend to change with time and will help in diverging the vehicle flow through the various routes of the road network. The road network problem is considered as a special case of Travelling Salesman Problem based combinatorial problem with changes and constraint imposed and also the steps of the algorithm is also changed to suit and quicken the solution finding process and imbibe the theory of chance and rejection subsequently. The results of application of the algorithm on the road network and its comparison with Ant Colony Optimization Algorithm & Intelligent Water Drops Algorithm show that the algorithm works well and provides the scope of utilization in similar kind of problems like path planning, scheduling, routing, and other constraint driven problems. EVOA is one of the very few algorithms which are readily applicable for discrete domain problems.

C. Sur (✉) · A. Shukla
Soft Computing and Expert System Laboratory, ABV- Indian Institute of Information Technology & Management, Gwalior, India
e-mail: chiranjibsur@gmail.com; dranupamshukla@gmail.com

9.1 Introduction

Road network management has become very challenging with the increase in the number of vehicles, inextensible infrastructure, lack of knowledge, congestion and lastly everyone tries to optimize things on their own and lands up with increased delay and heavy pollution. Hence with the effective implementation of the elaborated high speed mobile network, all-time connectivity and smart light weight applications there is requirement where the vehicle will not simply be guided but guided optimally so that the overall time, cost and effort spent is minimized, the scope of accidents and congestions is minimized and throughput of the network is enhanced. For this the requirements are: information collection system which can sent the information, a processing unit and lastly an interface for query or public interaction. So in this work we have provided a new approach for the path planning or routing of the graph scenarios. This new approach is another bio-inspired computation technique called EVOA which is better than the Ant Colony Optimization (ACO) and Intelligent Water Drop (IWD) with respect to the convergence rate for the same number of agent based implementation and thus is very suitable for the road network kind of scenario where the parameters of the topology changes rapidly and the validity of the solutions is of limited interval. The life style of the Egyptian Vulture has been a curiosity and field of study of many researchers because of its attitude, adaptive features, unique characteristics and enhanced skills and techniques for leading the lifestyle. Egyptian Vulture has been studied thoroughly and preventive measures are sought out for many decades ever since there occurred constant decrease in their numbers with some of their species became extinct with time. In this work we have formulate the activities of the Egyptian Vulture that can be applied for solution of real time problems and mobilized the process into some simple steps which can derive combinatorial optimization of graph based problems. There are some issues like fitness function, assumptions for choice of the options, etc. which need to be addressed (Sect. 9.2.4) for the benefit of the implementation of the algorithm. In the literature there are many nature inspired algorithms which has been proven to be highly efficient in optimization but only in the continuous domain and related fields where the application can be framed into equations and all the variables accept values continuously within a range. But a limited of them like ACO [22, 27], IWD [16, 30] can traverse a graph network for search based optimization. There are also some virtual transform of algorithm for discrete forms like discrete GA [24] and discrete PSO [23] which have limited utility for the real life problems. The main constituents of the evolutionary algorithm family are GA [3], PSO [3], cuckoo search [7], bat algorithm [8], harmony search [12], artificial immune system [11], simulated annealing [9], league championship algorithm [6], weed optimization [26], differential evolution [10], glowworm optimization [13], bacteria foraging algorithm [25] etc. where majority constitute equation based variation for local search and unable to handle sequence based problems. So there was requirement for development of efficient and specialized randomized meta-heuristics which can handle discrete problems. The development of the algorithm was primarily meant for the graph search problems through randomized selection and mixing up of the solution sets for opportunistic solution derivation amidst constraints imposed time to time by the various application system on which it is applied. Such a system is the road network system which is obsessed with heavy traffic and requires some efficient and optimized path finding mechanisms for enhanced management and avoidance of congestion.

The paper is arranged as Sect. 9.2 with the optimization algorithm, Sect. 9.3 puts up brief of road network used in simulation, Sect. 9.4 provides the steps of algorithm for optimal route search, Sect. 9.5 has computational results and graphs, Sect. 9.6 concludes with future work.

9.2 Egyptian Vulture Optimization Algorithm

The following subsections will gradually take up the various steps of the EVOA [28, 29] Meta-Heuristics as shown in Fig. 9.1. The two main activities of the Egyptian Vulture, which are considered here, are the tossing of pebbles and the ability of rolling things with twigs.

The path planning for road network should always start form the initialized datasets as the EVOA gradually searches for the path starting form initial and there is no scope of recombination of the intermediate nodes which can produce a new combination of the path. Also the number of nodes in solution varies and hence EVOA produces a self-adaptive feature to cope up with the problem.

9.2.1 Pebble Tossing

The Egyptian Vulture uses the pebbles for breakage of the eggs of the other birds which produce relatively harder eggs and only after breakage they can have the food inside. Two or three Egyptian Vulture continuously toss the pebbles on the egg with force until they break and they try to find the weak points or the crack points for success. This approach is used in this meta-heuristics for introduction of new solution in the solution set randomly at certain positions and hypothetically the solution breaks into the set and may bring about four possibilities depending upon probability and the two generated parameters for execution of the operations and selection of extension of the performance. Figure 9.2 provides the pictorial view of the pebble tossing step of the Egyptian Vulture. The two variables for the

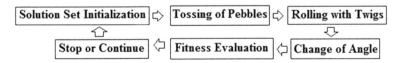

Fig. 9.1 Steps for Egyptian vulture optimization algorithm

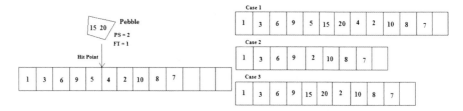

Fig. 9.2 Pictorial view of pebble tossing

determination of the extent of the operation are: PS = Pebble Size (level of occupy) where PS $\geq$ 0 and FT = Force of Tossing (level of removal) where FT $\geq$ 0. Hence, If PS > 0 Then "Get In" Else "No Get In". Also If FT > 0 Then "Removal" Else "No Removal" where "Get In" denotes occupancy and "Removal" is for removing. Now the Level of occupy denotes how many solutions should the pebble carry and must intrude forcefully into the solution set. Level of removal implies how many solutions are removed from the solution set. Both are generated randomly within a certain limit and the pebbles carrying PS number of nodes are also generated randomly considering that the combination can produce new solution set. Now FT denotes the number of nodes that are removed from either side of the point of hitting. Overall there are three combinations of operations are possible and are: Case 1: Get In & No Removal, Case 2: No Get In & No Removal, Case 3: Get In & Removal. Another criterion is the point of application of the step. Point of hitting is another criterion which requires attention and strategy must be determined for quickening the solution convergence process. For path finding problem the best strategy is considered if the discontinuous positions are targeted for application.

In Fig. 9.2 the path 1, 3, 6, 9, 5 are continuous and the rest from node 4 are not organized, then this point after 5 is considered for the operation to take place.

9.2.2 Rolling with Twigs

The rolling with twigs is another astonishing skill of the Egyptian Vulture with which they can roll an object for the purpose of movement or may be to perform other activity like finding the position or weak points or just giving a look over the other part which is facing the floor. This is perhaps the inherited skill of any bird for finding the right stick for any object they are trying to create or execute. Several birds have given testimony of making high quality nests during laying eggs. Such selection of sticks is mainly made for making the nest or positioning the right bend of the stick at the right place. Even some birds have given evidence of sewing the soft twigs with their beak. This activity of the Egyptian Vulture is considered as rolling of the solution set for changing of the positions of the variables to change the meaning and thus may create new solutions which may produce better fitness value and also better path when it comes for multi-objective optimization. Also when the hit points are less and the numbers of options are more, it may take a long time for

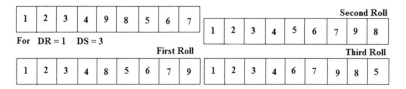

Fig. 9.3 Pictorial view of rolling with twigs for DS = 3, DR = 1

Say (1,2,3,4) forms a link, (7,6,5) another link, (8,9,10) another one. But there is no link between 4,7 and 5,8. But Change of Angle reverses the link 7,6,5 and tries to see if link exists between 4,5 or 7,8 or both.

The changed String can be the following, if links exist

Fig. 9.4 Pictorial view of change of angle

convergence to take place when appropriate matching of the random event to occur. For the "Rolling with Twigs" two more parametric variables are required which will direct the mathematical formulation of the event and also guide the implementation of the step. These two criteria for the determination of the extent of the operation are: DS = Degree of Roll where DS $\geq$ 0 denoting number of rolls. DR as Direction of Rolling where probabilistically we have: DR = 0 (for Right Rolling/Shift) or 1 (for Left Rolling/Shift), where 0 and 1 is generated randomly.

In Fig. 9.3 if we consider the numerical precedence of the nodes then we will find that the DS = 3, DR = 1 scheme will produce the required sequence in the second roll of the last five nodes (because the first four already makes a path) and will be accepted. Third roll will perform but the secondary fitness will reject it.

9.2.3 Change of Angle

This is another operation that the Egyptian Vulture can perform which derives its analogy from the change of angle of the tossing of pebbles so as to experiment with procedure and increase the chance of breakage of the hard eggs. Now the change of the angle is represented as a mutation step where the unconnected linked node sequence are reversed for the expectation of being connected and thus complete the sequence of nodes. Figure 9.4 gives a demonstration of such a step. This step is not permanent and is incorporated if only the path is improved.

This Change of Angle step can be multi-point step and the local search decides the points, number of nodes to be considered and depends on the number of nodes the string is holding. If the string is holding too many nodes and Pebble Tossing step cannot be performed then this step is a good option for local search and trying to figure the full path out of it.

9.2.4 Brief Description of Fitness Function

The fitness function is of utmost importance when it comes for the decision making of the system and optimization selection, but it is noticed that in majority graph based problems, obsessed with multi-objective optimization, that the complete path is reached after a number of iterations and by the mean time it is very difficult to clearly demarcate the better incomplete result from the others and it is in this case the act of probabilistic steps can worsen the solution. Also the acts of the operations need to be operated in proper places mainly on the node gaps where there is yet to make any linkage. Hence a brief description of the secondary fitness function needs to be addressed. The technique used in the simulation finds the linked consecutive nodes and is numbered with a number which denotes how many nodes are linked together at that portion. Then the secondary fitness is calculated as (summation of the fitness)/(number of nodes). High secondary fitness denotes that more numbers of nodes are linked together as a unit than the other solution string. But for the TSP as there occurs a link between every node, the secondary fitness will always be constant and will be of no use.

9.3 Model of Road Network Simulation

The EVOA is being applied on a road network for development of its efficiency and proper management of vehicles. The road network graph [27] is shown in Fig. 9.5. The network possess several paths and each of the edges has distance and average

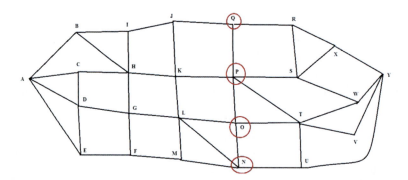

Fig. 9.5 Road network graph used for simulation

waiting time as its parameters and the overall fitness is calculated by the summation of the parameters of the edges of the derived path. The fitness function is kept simple and non-linear non-weighted sum of traveling time and waiting time. The main purpose of the EVO algorithm is finding the best path for the vehicles and analyzes the convergence rate with respect to Ant Colony Optimization. Convergence rate is important as quick finding is absolute necessary for this kind of highly dynamic systems.

However there are various assumptions considered for the simplicity of the road network and its vehicles and all agents and calculations are regarded as ideal. The vehicles are considered as of uniform dimension and capable of moving with non-accelerating velocity and thus the unnecessary minute details of the vehicle movement and also the minute variations are abstracted from the calculation. In the road graphs, it is considered that four points, which are the most vital points, the number of vehicles passing through these provides a rough estimations of how the flow of vehicles are distributed in the network. Another important assumption is the change in the dynamic parameter of average waiting time, which is in constant variation with respect to ranged pseudo-random generator. It is assumed that the presence of other vehicles in the road network produces an overall change in the waiting time and is considered as approximate average waiting time.

9.4 EVOA Methods for Optimized Route Search

In this section we have given details of the EVO algorithm for optimized route planning of the road network that can be utilized for traffic dispersion and management through guiding the vehicles to their destination following the best path available which minimizes the overall time spent from source to destination. The data structure mainly deals with the intermediate nodes with permanent source and destination.

Step 1: Initialize the Road Graph matrix $G = (V, E)$ and Road parameter matrix for each edge of the graph.

Step 2: Initialize N Solution String/Vector with x<<n nodes where n is the maximum possible nodes that a string can hold and x is the number of initial randomly generated nodes for a certain solution. Also initialize the unit bit string marker with 0 (for not complete) and can be 1 (when path is complete). This string marker will guide which of the solutions are complete and should not be altered anymore.

Step 3: Initialize the primary fitness and secondary fitness matrix.

Step 4: Prevent Duplicate nodes and Evaluate the secondary fitness of the initial strings (if any)

Step 5: Perform Tossing of Pebbles operation at points (or random points of the string mainly on the unstructured part of the solution). This depends upon implementation or probability of selection of the portion that require structuring.

The operation point initially ranged for the whole solution range but gradually the left limit of the range moves right as the initial path continuity is created.

Step 6: Perform Rolling of Twigs operation on selected (or the whole string if required) depending on the pseudorandom generation of the two operator parameters.

Step 7: Perform Change of Angle operation through selective reversal of solution subset. (This is some kind of extra effort introduced by the bird for efficient result and is a mutation operator)

Step 8: Prevent Duplicate nodes and Evaluate Secondary Fitness of each string of solution.

Step 9: If the path is found complete then mark the string marker as 1 and no more operation is performed on it.

Step 10: Evaluate Primary Fitness of each complete string set of solution or with string marker as 1.

Step 11: When the numbers of solutions are greater than a certain percentage of the total number of strings then Update the Global Best result with the best complete path depending upon the Primary Fitness.

Step 12: Check Condition for stopping or Start from Step 2. [As the system is dynamic the next iteration must start from the initialization]

Step 13: If iteration is complete provide the Best path for the vehicles for guidance.

Note: Duplicate node prevention (& Evaluate Secondary Fitness – if implementation logic depends on secondary fitness) step operation is required to be executed after each operation to prevent confusion and enhance the quality of the solution.

9.5 Computational Results

The following are the graphs generated for the road network simulation with the EVOA algorithm.

Figures 9.6, 9.7, 9.8, 9.9 and 9.10 have provided the comparison of the simulation for the global best, cumulative global best & average cumulative global best for optimized path with respect to the total time denoted as the sum of travel time and waiting time (total time = travel time + waiting time), and travel time and waiting time individually for ten agent based simulation and it is found that EVO has better convergence rate than the ACO. Travel time and waiting time are individual parameters for each edge of the network graph. The more number of agents will scope for more exploration but will increase the simulation time proportionally. The main reasons for the enhanced convergence rate of EVO than ACO are:

- In ACO if the path formed form a loop, the agent is wasted. But in EVO the scheme of dropping the nodes will revive the agent. Intelligent computation must control the inclusion and exclusion of nodes for the path.
- In ACO (as Mean-Minded ACO [27]) the agents are influenced by the pheromone level of the paths probabilistically and more concentrated on the better

9 Road Traffic Management Using Egyptian Vulture Optimization Algorithm...   115

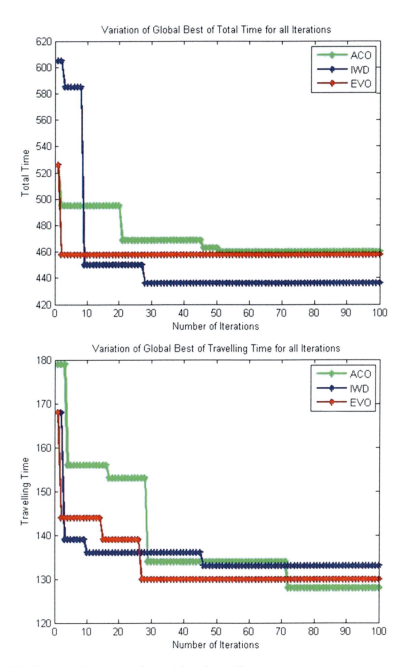

Fig. 9.6 Variation of global best for total time & travelling time

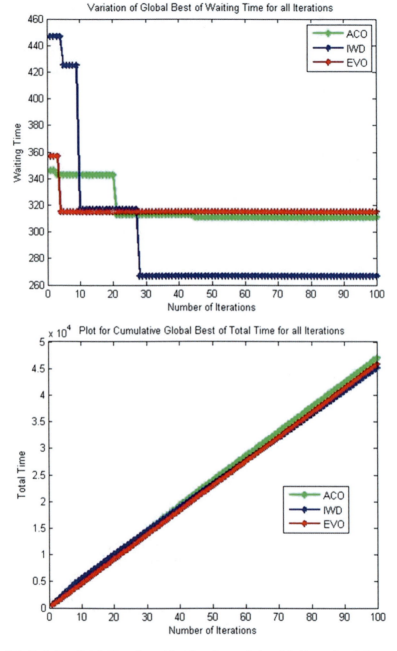

Fig. 9.7 Variation of global best for waiting time & cumulative global best of total time

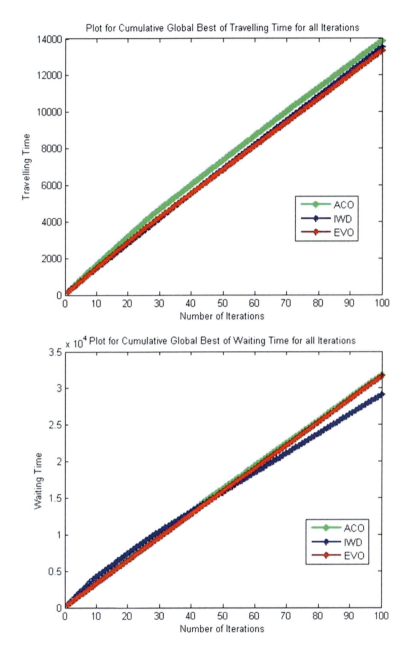

Fig. 9.8 Variation of cumulative global best of travelling time & waiting time

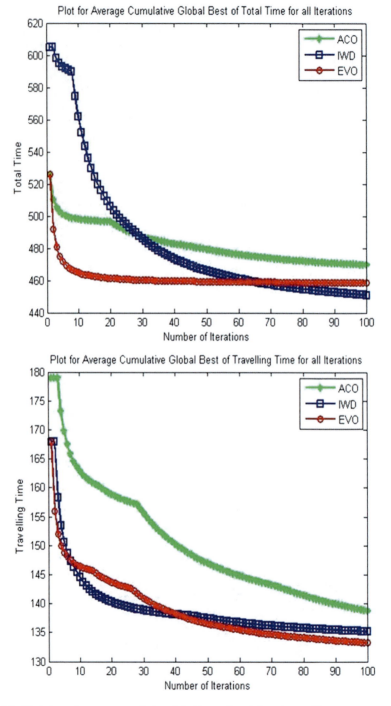

Fig. 9.9 Variation of average cumulative global best of total time & travelling time

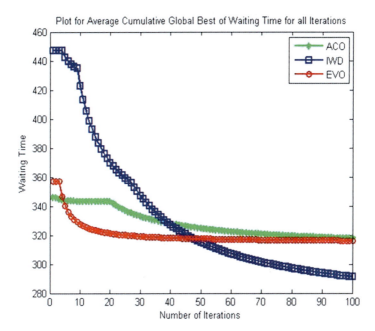

Fig. 9.10 Variation of global best for waiting time

average better path for the individual agent, but in EVO pure exploration, with new combination generation, takes place.

- In EVO, all paths are formed and thus the scope of better path finding or formation is more than the ACO, where agents are influenced by the pseudo communication of the other ant agents. Also the later iterations are influenced by the previous iterations. In dynamic situations like the traffic level of the road network, ACO should always start from scratch but this will affect the unique principle of the algorithm and will hamper its performance. However the decision making of choice of path in ACO is sometimes lend to probability for exploration. Thus in dynamic systems, the previous account for better paths are not utilized in the new iteration start as the parameters of the system has changed and moreover there has appeared a new topology. Also in the same iteration, the already passed by agents are not having the benefit of the pheromone deposited by the other agents.

The uniqueness of the algorithm lies in the randomness of node selection, adaptability and rearrangement process. The solution is not created dynamically like in ACO and IWD, but are added then tried to link up through the process of random node selection and guided replacement, and another is rearrangement (through Rolling with twigs & Change of angle). Also two important characteristics of the algorithm are: The algorithm is also unique and also found experimentally for the production of different path solutions for low number of solution agents. Another is that if the numbers of vulture agent increase then the process of solution generation is enhanced. Multi-agent vulture applied on the same solution will decrease the chance of rejection of nodes for a solution set. However random selection must constitute all the nodes with equal probability. Otherwise another problem is encountered for the algorithm is during the completion stage where there are required of only one or two node and due to randomness they don't appear. As a result many solutions may go wasted. At this stage the requirement is to loop the whole node set for the solution set and completion of the solution will be enhanced. The plots are mainly simulated for total time and the subsequent best entries for waiting time and travelling time are taken individually and hence the plots are not the best but are competitive and acceptable as a proof of the performance of the Egyptian Vulture Optimization Algorithm. The agents are multiple in numbers but the EVOA cannot be called as a swarm as the agents are not dependent on each other for any mean.

9.6 Conclusion and Future Works

The results through graphs shown in the previous Section clearly reflected the potential of the Egyptian Vulture Optimization Algorithm as a better discrete heuristics and in handling graph based combinatorial problems and performing quite well for constraint imposed scenario example of road network where it has been worked on weighted multi-objective optimization. The new meta-heuristics provides some probabilistic approach on some local search and analysis of secondary fitness which is operated on the incomplete path solution. Also this new Egyptian Vulture Optimization Algorithm provides quite a number of paths in the solution domain and out of which the best one derived is declared as the global best for that iteration. Also a least path derivation of the road network is plotted, the closeness of the least path solution it has been able derived out of the new meta-heuristic search and also its dependency on the number of parallel solution strings it has considered.

Lot of work can be extended for the new heuristic for other kind of combinatorial optimization applications and requires proper handling of the heuristic steps according to the problem and constraints. The current approach is biased for graph based problems and lacks a local search strategy like that of influencing change in the value of the variable in Particle Swarm Optimization Algorithm and Bacteria Foraging Optimization Algorithm. Also in graph based problems such local search strategy are either not applicable or not required. The algorithm can be introduced with such kind small step extension in any direction for quick convergence of the required minima or maxima.

References

1. http://en.wikipedia.org/wiki/Egyptian_Vulture
2. http://www.flickr.com/photos/spangles44/5600556141/
3. C. Blum, A. Roli, Metaheuristics in combinatorial optimization: overview and conceptual comparison. ACM Comput. Surv. **35**(3), 268–308 (2003)
4. J. Kennedy, R. Eberhart, Particle swarm optimization, *Neural Networks, 1995. Proceedings, IEEE International Conference on*, vol. 4, 1995, pp. 1942–1948
5. D. Karaboga, An idea based on honey bee swarm for numerical optimization. Technical Report TR06, Erciyes University, Oct 2005
6. Ali Husseinzadeh Kashan, League Championship Algorithm: A New Algorithm for Numerical Function Optimization, in *Proceedings of the 2009 International Conference of Soft Computing and Pattern Recognition (SOCPAR '09)* (IEEE Computer Society, Washington, DC, 2009), pp. 43–48
7. X.-S. Yang, S. Deb, Cuckoo search via Levy flights, in *World Congress on Nature & Biologically Inspired Computing (NaBIC 2009)*. IEEE Publication, pp. 210–214
8. X.-S. Yang, A new metaheuristic bat-inspired algorithm, in: Nature Inspired Cooperative Strategies for Optimization (NICSO 2010), Science **284**, 65–74 (2010)
9. S. Kirkpatrick, C.D. Gelatt Jr., M.P. Vecchi, Optimization by simulated annealing. Science **220**(4598), 671–680 (1983)
10. R. Storn, K. Price, Differential evolution – a simple and efficient heuristic for global optimization over continuous spaces. J. Global Optim. **11**(4), 341–359 (1997)
11. J.D. Farmer, N. Packard, A. Perelson, The immune system, adaptation and machine learning. Physica D **22**(1–3), 187–204 (1986)
12. Z.W. Geem, J.H. Kim, G.V. Loganathan, A new heuristic optimization algorithm: harmony search. Simulation **76**(2), 60–68 (2001)
13. K. Krishnanand, D. Ghose, Glowworm swarm optimization for simultaneous capture of multiple local optima of multimodal functions. Swarm Intell. **3**(2), 87–124 (2009)
14. O.B. Haddad, A. Afshar, M.A. Mariño et al., Honey-bees mating optimization (HBMO) algorithm: a new heuristic approach for water resources optimization. Water Resour. Manage. **20**(5), 661–680 (2006)
15. K. Tamura, K. Yasuda, Primary study of spiral dynamics inspired optimization. IEEJ Trans. Electr. Electron. Eng. **6**(S1), S98–S100 (2011)
16. H. Shah-Hosseini, The intelligent water drops algorithm: a nature-inspired swarm-based optimization algorithm. Int. J. Bio-Inspired Comput. **1**(1/2), 71–79 (2009)
17. P. Civicioglu, Transforming geocentric cartesian coordinates to geodetic coordinates by using differential search algorithm. Comput. Geosci. **46**, 229–247 (2012)
18. M.H. Tayarani-N, M.R. Akbarzadeh-T, Magnetic Optimization Algorithms a new synthesis, *Evolutionary Computation, 2008. CEC 2008. (IEEE World Congress on Computational Intelligence). IEEE Congress on*, 1–6 June 2008, pp. 2659–2664
19. C.W. Reynolds, Flocks, herds and schools: a distributed behavioral model. Comput. Graphic. **21**(4), 25–34 (1987)
20. A. Kaveh, S. Talatahari, A novel heuristic optimization method: charged system search. Acta Mech. **213**(3–4), 267–289 (2010)
21. A.H. Gandomi, A.H. Alavi, Krill Herd algorithm: a new bio-inspired optimization algorithm. Commun. Nonlinear Sci. Numer. Simulat. (2012)
22. M. Dorigo, L.M. Gambardella, Ant colony system: a cooperative learning approach to the traveling salesman problem. IEEE Trans. Evol. Comput. **1**(1), 53–66 (1997)
23. J. Kennedy, R.C. Eberhart, A discrete version of the particle swarm algorithm, *Proceedings of Conference on Systems, Man, and Cybernetics* (IEEE Services Center, NJ, 1997), pp. 4104–4108

24. Pengfei Guo, Xuezhi Wang, Yingshi Han, A hybrid genetic algorithm for structural optimization with discrete variables, *Internet Computing & Information Services (ICICIS), 2011 International Conference on*, 17–18 Sept 2011, pp. 223–226
25. S. Das, A. Biswas, S. Dasgupta, A. Abraham, Bacterial foraging optimization algorithm theoretical foundations, analysis, and applications, in *Foundations of Computational Intelligence Volume 3: Global Optimization* (Springer, Berlin/Heidelberg, 2009), pp. 23–55
26. A.R. Mehrabian, C. Lucas, A novel numerical optimization algorithm inspired from weed colonization. Ecol. Inform. **1**, 355–366 (2006)
27. C. Sur, S. Sharma, A. Shukla, Analysis & modeling multi-breeded Mean-Minded ant colony optimization of agent based Road Vehicle Routing Management, *Internet Technology and Secured Transactions, 2012 International Conference For*, Dec 2012, pp. 634–641
28. C. Sur, S. Sharma, A. Shukla, Egyptian Vulture Optimization Algorithm – A New Nature Inspired Meta-heuristics for Knapsack Problem, *The 9th International Conference on Computing and Information Technology (IC2IT2013)*, vol. 209, 2013, pp. 227–237
29. C. Sur, S. Sharma, A. Shukla, Solving Travelling Salesman Problem Using Egyptian Vulture Optimization Algorithm – A New Approach, *20th International Conference on Intelligent Information Systems*, vol. 7912 (IIS 2013), 2013, pp. 254–267
30. C. Sur, S. Sharma, A. Shukla, Multi-objective adaptive intelligent water drops algorithm for optimization & vehicle guidance in road graph network, *Informatics, Electronics & Vision (ICIEV), 2013 International Conference on*, 17–18 May 2013, pp. 1,6

Part III
Ad Hoc and Sensor Networks

Chapter 10
A Novel Bloom Filter Based Variant of Elliptic Curve Digital Signature Algorithm for Wireless Sensor Networks

Vivaksha Jariwala, Prafulla Kumar, and Devesh C. Jinwala

Abstract In this paper, our focus is on the investigation for further improvement (with its variants already existing) upon the Elliptic Curve Digital Signature Algorithm (ECDSA). The security of ECDSA is based on the Elliptic Curve Discrete Logarithm Problem. Though, ECDSA uses the same number to generate two separate signatures as per the original protocol, it is possible for an adversary to forge the signature. There have been number of improvements proposed to circumvent the issue. However, we propose here a probabilistic and improved bloom filter based variant of ECDSA that while being optimal enhances the security strength of ECDSA. With the theoretical analysis supplemented with our experimentation on the TinyOS platform, we show that it is appropriate to employ in the resource-constrained environment of Wireless Senor Networks (WSNs).

10.1 Introduction

Devising the security protocols in the resource constrained Wireless Sensor Networks (WSNs) environment is nontrivial. This is principally due to the (1) ubiquitous and pervasive deployment of the sensor nodes, (2) the resource intensive nature of security algorithms and (3) the severe constraints in memory, computational and communication resources in WSNs [1].

V. Jariwala
C.K. Pithawalla College of Engineering and Technology,
Computer Engineering Department, Surat, India
e-mail: vivakshajariwala@gmail.com

P. Kumar
Department of Information Technology, Govt. of India, Delhi, India

D.C. Jinwala (✉)
S.V. National Institute of Technology,
Computer Engineering Department, Surat, India
e-mail: dcjinwala@gmail.com

Any security algorithm for WSNs, must minimally provide the attributes viz. confidentiality, data integrity and entity authentication. However, a protocol that provides confidentiality without the support for data integrity is meaningless. Hence, we focus on investigating the techniques for supporting data integrity. As per our literature survey, we categorize the techniques for supporting data integrity in WSNs into three classes viz. signature based, hash function based and Message Authentication Code (MAC) based. Observing the fact that the digital signature based approach yields non-repudiation property, our focus is on digital signature based solution for supporting data integrity.

We realize that there are two types of ECC based signature available in the literature viz. Elliptic Curve Digital Signature Algorithm (ECDSA) [2] and Elliptic Curve Pintsov Vanstone Signature (ECPVS) that provide additional confidentiality features [3]. ECDSA in TinyECC [4] is an example of a standard signature based algorithm for data integrity – the primary focus in this research. The ECDSA has a smaller key size that leads to the faster computation time and reduction in processing power, storage space and bandwidth. This makes the ECDSA implementation, suitable for the resource-constrained environments such as the WSNs.

ECDSA uses a random number k to generate the signature. Therefore, by any chance, if the same number is used for generating another signature, then an adversary can find the value of private key x. The adversary can then use it to generate another signature. The obvious solution is to employ a random number generator that assures non-repetition of its output. However, in that case, the strength of ECDSA algorithm is not *intrinsic* i.e. not built into the algorithm, but dependent on the implementation. One of the panaceas to this issue is to use two different numbers viz. k_1 and k_2 for signature generation [5]. As the authors prove formally in their paper, this indeed increases the security strength. Indeed, numerous efforts exist in literature that focuses on improving the ECDSA algorithm itself. However, the issue that crops up is why not to use three different numbers – especially if using two numbers k_1 and k_2 instead of a single random number k increases the security strength, then using one more shall surely further increase the same. We indeed experiment with it as explained further in Sect. 10.4.

However, two vital issues that crop up here are as follows: (1) if increasing the number of random numbers used to generate the signature increases the security strength, then why not to employ still higher numbers of such random numbers. (2) the resulting overhead in generating increased number of random numbers is tolerable? Such issues provoked us to explore any other alternative to argue that eventually the goal in using multiple k_i's is only to prevent the probability of their repeated occurrences so that the signature cannot be forged. Hence, is it not worthwhile to explore any alternate mechanism to ensure increased security strength of ECDSA while at the same time preventing multiple invocation of k_i's ? With this aspect in focus, and with the resource constraints in WSNs at the backdrop, we observed that the space-efficient probabilistic set membership test data structure viz. bloom filter can be employed for the purpose here [6]. Thus, we propose a bloom filter based ECDSA that uses set membership test methods and light-weight hash functions [7] to generate unique secret every time. Thus, because each k_i generated is tested a priori for its use in earlier runs before actually using it, it is not possible for the adversary to use it maliciously to generate a false signature.

We use the TinyOS [8] environment for WSNs to empirically evaluate and benchmark the implementation of our approach against all the other previously proposed approaches. We implement variants of ECDSA viz. one using two different secrets [5], another using three different secrets, one more specifically designed for limited capacity signer [9] and another specifically designed for limited capacity [9] verifier. We also compare the overhead in all these implementations using *RAM requirements, ROM requirements* and *energy consumption* as the metrics. As our evaluation also shows, the overhead in our approach is tolerable even in the resource-constrained environments while imparting the necessary security strength to the ECDSA. There is no doubt that our approach being simple; extension of our approach to the other environments (beyond WSNs) is also possible.

To the best of our knowledge, our proposal is a simple and yet a unique attempt that applies the set membership test operation data structure viz. bloom filter to an advantage to the ECDSA in the resource constrained environment of WSNs while enhancing the security of ECDSA.

The remainder of this paper is organized as follows. Section 10.2 gives a detailed description of original ECDSA scheme and its security. Section 10.3 discusses various variant of ECDSA and its security. Section 10.4 describes our proposed variant of ECDSA based on bloom filter. Section 10.5 describes our results and analysis followed by conclusion in Sect. 10.6.

10.2 Elliptic Curve Digital Signature Algorithm (ECDSA)

The Elliptic Curve Digital Signature Algorithm consists of three phases (1) Key generation. (2) Signature generation and (3) Signature verification. The domain parameter for the ECDSA consists of a suitably chosen elliptic curve E defined over a finite field Fp of characteristic p, and a base point G {∈} Ep(a,b) with order n. Thus, a typical setup phase of ECDSA would appear as follows [2]:

- Select a random or pseudo-random integer x such that $1 \leq x \leq n - 1$.
- Compute $Q = xG$.
- A's public key is Q; A's private key is x.

To sign a message m, an entity A with domain parameters (p,Ep(a,b),G,n) and associated key pair (x,Q) does the following [2]:

```
1. Select an integer k such that 1≤k≤n-1.
2. Compute kQ=(x₁,y₁).
3. Compute r=x₁(mod n). If r=0 then go to step 1.
4. Compute k⁻¹(mod n).
```

5. Compute $SHA^{-1}(m)$ and convert this string to an integer H (m)
6. Compute $s=k^{-1}(H(m)+xr) \pmod{n}$.
7. If s=0, then go to step 1.
8. A's signature for the message m is (r,s).

To verify A's signature (r, s) on m, B obtains an authentic copy of A's domain parameter (p,Ep(a,b),G,n) and associated public key Q. B then does the following [2]:

1. Compute SHA-1(m) and convert this string to an integer H (m).
2. Compute $w=s^{-1} \pmod{n}$.
3. Compute $u_1=H(m)w \pmod{n}$ and $u_2=rw \pmod{n}$.
4. Compute $X=(x_2,y_2)=u_1G+u_2Q$.
5. If X=O, then reject the signature. Otherwise, compute $v=x_2 \pmod{n}$.
6. Accept the signature if and only if v=r.

10.2.1 A Possible Attack on ECDSA

In the original ECDSA, the integer k should unique to sign distinct messages. This means that every time a message is to be signed, a uniquely distinct secret k should be used to sign message. If it is not so, the private key x can be disclosed, thus making the scheme vulnerable to the attacks. This is illustrated in [5] as discussed below:

Assume that same secret k is used to generate two ECDSA signature (r,s_1) and (r,s_2) for two different messages m_1 and m_2. Thus,

$$s_1 = k^{-1}(H(m_1) + xr)(mod\ n) \text{ and } s_2 = k^{-1}(H(m_2) + xr)(mod\ n)$$

Here $H(m_1)$ is Hash of message m_1 and $H(m_2)$ is Hash of message m_2.

$$ks_1 = H(m_1) + xr \pmod{n} \quad (10.1)$$
$$ks_2 = H(m_2) + xr \pmod{n} \quad (10.2)$$

Now, if we subtract Eq. 10.1 from Eq. 10.2

$$k(s_2 - s_1) = H(m_2) - H(m_1) \pmod{n}$$

If, $s_2 \neq s_1 \pmod{n}$, that occurs with high probability, then,

$$k = (s_2 - s_1)^{-1}(H(m_1) - H(m_2)) \pmod{n}$$

Hence, an adversary can determine k and then use k to reveal the secret x. Thus, if the same secret k is used to sign two different messages, adversary can easily reveal secret x. There are numerous attempts made in the literature to deal with this issue [5, 9–12]. We critically analyze these efforts and use the same analysis to justify our motivation in proposing a new variant of ECDSA.

10.3 Variants of ECDSA

10.3.1 Variant: Limited Computation Capacity Signer

Authors in [9] proposed variant for limited computation capacity signer. The scheme in [9] is suitable for limited computation capability like a signer using his smart card that stores secret key and signs a message on a terminal. Here the advantage is that there is no need of calculating inverse of x in each individual signing operation. X is the private key of the signer that will remain stable for a period, it can be pre computed and stored in the key generation phase itself. Here in this scheme [9] also attack possible on basic ECDSA is feasible. Hence, it is also not secure.

10.3.2 Variant: Limited Computation Capacity Verifier

Authors in [9] proposed variant for limited computation capacity verifier. The scheme proposed in [9] is suitable for the verifier who has limited compute apparatus. That is, in this scheme the complexity of verification operation is lesser as compared to that of the previous schemes. In this scheme, k^{-1} is no longer be calculated, but we must calculate $(h + rx)^{-1}$ in the signing phase. However, there is no need of calculating inverse in verification phase that is one of the most expensive operations in modular arithmetic. Therefore, the complexity of the verification process is less in this scheme. Here in this scheme [9] also attack possible on basic ECDSA is feasible. Hence, it is also not secure.

10.3.3 Variant: Using Two Different Secrets

In the variant proposed in [5], the digital signatures are generated using two secrets k_1 and k_2 instead of relying upon a single secret k. Here, x cannot be determined even if the same secret k_1 and k_2 is repeated. The authors in their paper formally prove this:

If a signature (r_1, s) on a message m was indeed generated by A, then $s = k^{-1}{}_1 (H(m)k_2 + x(r_1 + r_2))$ (mod n). If the same secret k_1, k_2 was used to generate ECDSA signatures (r_1, s_1) and (r_1, s_2) on two different messages m_1 and m_2. Then [5],

$$s_1 = k^{-1}{}_1(H(m_1)k_2 + x(r_1 + r_2)) \text{ (mod n) and}$$
$$s_2 = k^{-1}{}_1(H(m_2)k_2 + x(r_1 + r_2)) \text{ (mod n)}$$

Where $H(m_1) = SHA - 1(m_1)$ and $H(m_2) = SHA - 1(m_2)$. Then

$$k_1 s_1 = H(m_1)k_2 + x(r_1 + r_2) \text{(mod n)} \qquad (10.3)$$
$$k_1 s_2 = H(m_2)k_2 + x(r_1 + r_2) \text{(mod n)} \qquad (10.4)$$

Subtraction gives $k_1(s_1 - s_2) = (H(m_1) - H(m_2))k_2$(mod n). We cannot determine k by this equation and then use this to recover x. Hence, this scheme is more secure. However, the processes are more complex than the original ECDSA.

10.4 Proposed Variant Based on Bloom Filter

As already discussed in Sect. 10.2.1 if the same secret k is used to sign two different messages, an adversary can easily reveal the secret x and with secret x, any adversary can easily generate malicious signature on the message. To, surmount that we can use additional secrets to generate signature as proposed in [5]. We have also proposed a variant of ECDSA that uses multiple secrets (three) to generate signature. Our proposed variant of ECDSA using multiple secrets is as follows.

Signature Generation:

1. Select k_1, k_2, k_3 $1 \leq k_1, k_2, k_3 < n$.
2. $k_1 G = (x_1, y_1)$, $r_1 = x_1$ (mod n)
3. $k_2 G = (x_2, y_2)$, $r_2 = x_2$ (mod n)
4. $k_3 G = (x_3, y_3)$, $r_3 = x_3$ (mod n)
5. $s = k_1^{-1} (H(m) k_2 k_3 + x (r_1 + r_2 + r_3))$ (mod n)
6. (r_1, s) is the signature of m.

Signature Verification:

1. $w = s^{-1}$ (mod n)
2. $u_1 = H(m) w k_2 k_3$ (mod n)
3. $u_2 = (r_1 + r_2 + r_3) w$ (mod n)
4. $u_1 G + u_2 Q = (x_4, y_4)$,
5. $v = x_4$ (mod n)
6. $v = r_1 \rightarrow$ accept the signature

In this scheme, we cannot determine k_1, k_2, k_3 and then use this to recover x. Hence, this scheme is more secure. However, when using additional random numbers (secrets) to enhance the security strength, a vital issue that crops up is viz. how many such secrets invocations to use? To answer the same, we also propose a solution that uses the set membership data structure viz. bloom filter to avoid repetition of a random number (secrets) used in ECDSA.

Our proposed approach uses a bloom filter to generate unique k that can be used to sign different messages. Hence, our approach is using different secrets every time to generate different signatures without multiple secrets and consequently there are no chances of generation of false signature. The only argument against the usage of the proposed variant could be *if at all we have a strong random number generator that ensures non-repetition, is it necessary to employ this variant.* However, in that case, ECDSA remains dependent on the implementation to be secure – the algorithm lacks intrinsic security strength. Hence, our proposal is justified in that in enhances the intrinsic security strength of the ECDSA algorithm without assuming any guarantees from the underlying implementation.

Following subsection shows the description of bloom filter.

10.4.1 Bloom Filter

A Bloom filter [6], is a space-efficient probabilistic data structure that is used to test whether an element is a member of a set or not. This compressed representation is the payoff for allowing a small rate of false positives in membership queries; that is, queries might incorrectly know an element as member of the set. Consider a set $A = \{a_1, a_2, \ldots, a_n\}$ of n elements. Bloom filters describe membership information of A using a bit vector V of length m. For this, k hash functions, $h_1, h_2, \ldots, h_k$ with $h_i : X \rightarrow \{1..m\}$, are used as described below: The following procedure builds an m bits bloom filter, corresponding to a set A and using $h_1, h_2, \ldots, h_k$ hash functions [6]:

```
Procedure BloomFilter(set A, hash_functions, integer m)
   returns filter
filter=allocate m bits initialized to 0
for each a_i in A:
    for each hash function h_j:
filter[h_j(a_i)]=1
   end for each
end for each
return filter
```

Therefore, if a_i is member of a set A, in the resulting bloom filter all bits obtained corresponding to the hashed values of a_i are set to 1. Testing for membership of an element is equivalent to testing that all corresponding bits of bloom filter are set [6]:

```
Procedure MembershipTest (elm, filter, hash_functions)
returns yes/no
for each hash function h_j:
if filter[h_j(elm)] !=1 return No
end for each
return Yes
```

As new elements are added to the set, filters can be built incrementally. After that, the corresponding positions are computed through the small hash functions and bits are set in the filter. Moreover, the filter expressing the reunion of two sets is simply computed as the bit-wise OR applied over the two corresponding bloom filters.

10.4.2 Proposed Variant of ECDSA Using Bloom Filter

Select Ep(a,b), x, and $1 \leq x < n$. Select G $\in$ Ep(a,b) with order n and compute Q = xG. Public key: (Ep(a,b), p,G,n,Q). Private key: x.

Signature Generation:
To sign a message m, an entity A with domain parameters (p,Ep(a,b),G,n) and associated key pair (x,Q) does the following:

```
1. Select k, 1≤k<n.
2. Create Bloom Filter
3. Call MembershipTest for k
4. If returns yes go to step 1
5. Else go to next step
6. kG=(x_1,y_1), r=x_1 (mod n)
7. s=k^-1 (H(m)+xr) (mod n)
8. (r, s) is the signature of m.
```

Signature Verification:
The receiver can verify the authenticity of sender's signature (r,s) for message m by performing following operations.

1. w=s^{-1}(mod n)
2. u$_1$=H(m)w(mod n)
3. u$_2$=rw(mod n)
4. u$_1$G+u$_2$Q=(x$_2$,y$_2$),
5. v=x$_2$(mod n)
6. v=r accept the signature

10.4.3 Proof of the Scheme

Our proposed approach uses bloom filter to generate a unique k that can be used to sign different messages. Therefore, an attack possible on basic ECDSA is not possible on our variant of ECDSA. In our bloom filter based approach, we first call the set membership test. If number generated is there in the set then we generate another number, whereas if not, we use that secret for generation of signature. Therefore, every time different secrets are used to generate different signatures. For example, let secret bloom$_1$ and bloom$_2$ are used to generate two ECDSA signatures viz. (r, s$_1$) and (r, s$_2$) for two different messages m$_1$ and m$_2$. Thus,

$$s_1 = \text{bloom}_1^{-1}(H(m_1) + xr)(\text{mod } n) \text{ and } s_2 = \text{bloom}_2^{-1}(H(m_2) + xr)(\text{mod } n)$$

Here H(m$_1$) is Hash of message m$_1$ and H(m$_2$) is Hash of message m$_2$.

$$\text{bloom}_1 s_1 = H(m_1) + xr \ (\text{mod } n) \tag{10.5}$$

$$\text{bloom}_2 s_2 = H(m_2) + xr (\text{mod } n) \tag{10.6}$$

Now, if we subtract Eq. 10.5 from Eq. 10.6

$$(\text{bloom}_2 s_2 - \text{bloom}_1 s_1) = H(m_2) - H(m_1) \ (\text{mod } n)$$

If, $s_2 \neq s_1$(mod n), that occurs with high probability, then, we cannot determine bloom$_1$ and bloom$_2$ by this equation and then use this to recover x. Hence, this scheme is more secure.

10.5 Results and Analysis

Table 10.1 shows the comparison of various variants of ECDSA. From the table we can see that the variants that use two different secrets, three different secrets and the proposed one using bloom filter are not vulnerable to attack. However, instead of generating two separate secrets or three separate secrets for signature our approach a priori checks for the uniqueness of the secret. Therefore, our approach enhances

Table 10.1 Comparison of various variants of ECDSA

Algorithm	No. of secrets used	Signature generation	Signature verification	Attacks if same secrets are used
ECDSA	1	$s = k^{-1}(h + xr)$	$u_1 = hs^{-1}$ (mod n) $u_2 = rs^{-1}$ (mod n)	Vulnerable
ECDSA – limited capacity signer	1	$s = x^{-1}(rk - h)$	$u_1 = hr^{-1}$ (mod n) $u_2 = sr^{-1}$ (mod n)	Vulnerable
ECDSA – limited capacity verifier	1	$s = k(h + rx)^{-1}$	$u_1 = hs$ (mod n) $u_2 = rs$ (mod n)	Vulnerable
ECDSA – two secrets	2	$s = k^{-1}_1(hk_1 + x(r_1 + r_2))$	$u_1 = hs^{-1} k_2$ (mod n) $u_2 = (r_1 + r_2)s^{-1}$	Not vulnerable
ECDSA – three secrets	3	$s = k^{-1}_1(hk_1 k_2 + x(r_1 + r_{2+} r_3))$	$u_1 = hs^{-1} k_2 k_3$ (mod n) $u_2 = (r_1 + r_2 + r_3)s^{-1}$	Not vulnerable
Bloom filter based ECDSA	1	$s = k^{-1}(h + xr)$	$u_1 = hw$ (mod n) $u_2 = rw$ (mod n)	Not vulnerable

the security of basic ECDSA using a single secret only and increases intrinsic strength of algorithm.

In this, we attempt to add our application in TinyECC [4] of TinyOS [8]. In order to do this, we propose and implement our novel variant of ECDSA based on bloom filter and try to evaluate these all variants based on the different metrics viz. storage requirements (*RAM* and *ROM*) using TOSSIM [13] and energy in joule using Avrora [14]. Any algorithm for WSNs must be designed carefully to work in the resource-constrained environment. Hence, we use above-mentioned metrics that directly affect the lifetime of the sensor nodes to evaluate the performance of the proposed variant of ECDSA. In this section, we show our experimental results for various variants of ECDSA including our proposed variant based on the above-mentioned metrics.

Figure 10.1 shows RAM requirements for various variants of ECDSA. In the figure, ECDSA is the basic ECDSA, variant 1 is limited computation capacity signer, variant 2 is limited computation capacity verifier, variant 3 is using two different secrets, variant 4 is using three different secrets and last is our proposed bloom filter based variant. We can say that our approach using bloom filter requires only 6 % more RAM than that of other variant of ECDSA but that is at the cost of additional security.

Figure 10.2 shows ROM requirements for various variants of ECDSA. We can say that our approach using bloom filter requires only 2 % more ROM than that of other variant of ECDSA but that is at the cost of additional security.

Figure 10.3 shows Energy consumption for various variants of EC-DSA. We can say that our approach using bloom filter requires approximately same energy compare to other variant of ECDSA and that is at the cost of additional security.

10 A Novel Bloom Filter Based Variant of Elliptic Curve Digital Signature...

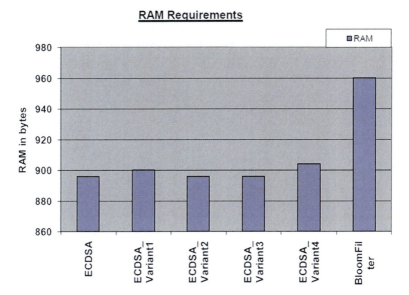

Fig. 10.1 RAM requirements of various variants of ECDSA

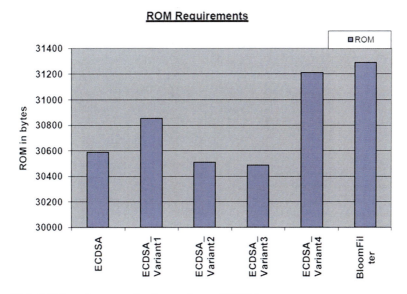

Fig. 10.2 ROM requirements of various variants of ECDSA

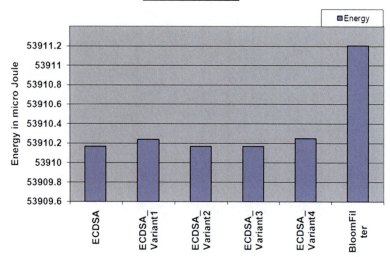

Fig. 10.3 Energy consumption of various variants of ECDSA

10.6 Conclusion

In this paper, we analyze the basic ECDSA and its various variants. In addition, we also propose our own bloom filter based variant of ECDSA that enhances the intrinsic security strength of the ECDSA algorithm. To justify that the variant can work feasibly in WSNs, we implement the proposed variant including all the other variants too and compare the same empirically using the metrics viz. *Storage (RAM, ROM)* and *Energy Consumption*. Our empirical evaluation and proof of the algorithm clearly shows that our variant of ECDSA increases intrinsic security of ECDSA and is suitable for any application demanding integrity support in resources constrained environment of WSNs.

Acknowledgements The work contained herein was carried with support from a sponsored project from the Department of Electronics and Information Technology, Ministry of Communications and Information Technology, Govt. of India. The authors remain grateful to the sponsoring agency for the same.

References

1. I.F. Akyildiz, W. Su, Y. Sankarasubramaniam, E. Cayirci, Wireless sensor networks: a survey. J. Comput. Netw. **38**(4), 393–422 (2002)
2. D. Johnson, A. Menezes, S. Vanstone, The elliptic curve digital signature algorithm (ECDSA). Int. J. Info. Secur. **1**(1), 36–63 (2001)
3. SEC 3, Standards for efficient cryptography, elliptic curve signatures giving partial message recovery, http://www.secg.org

4. A. Liu, P. Kampanakis, P. Ning, TinyECC: elliptic curve cryptography for sensor networks, in *Proceedings of International Conference on Information Processing in Sensor Networks*, St. Louis, MO, 2008, pp. 245–256
5. Hung-Zih Liao, Yuan-Yuan Shen, On the elliptic curve digital signature algorithm. Tunghai Sci. **8**, 109–126 (2006)
6. B. Bloom, Space/time trade-offs in hash coding with allowable errors. Commun. ACM **13**(7), 422–426 (1970)
7. J. Lawrence Carter, M.N. Wegman, Universal classes of hash functions. J. Comput. Syst. Sci. **18**, 143–154 (1979)
8. J. Hill et al., System architecture directions for networked sensors, in *Proceedings of the 9th International Conference on Architectural Support for Programming Languages and Operating Systems (ASPLOS 2000)*, ACM Press, New York, 2000, pp. 93–104
9. Hu Junru, The improved elliptic curve digital signature algorithm, in *Proceeding of International Conference on Electronic & Mechanical Engineering and Information Technology*, IEEE, Harbin, China, August 12–14, (2011)
10. A. Khalique, K. Singh, S. Sood, Implementation of elliptic curve digital signature algorithm. Int. J. Comput. Appl. **2**(2), 21–27 (2010)
11. Technical Guideline TR-03111 Elliptic Curve Cryptography Version 2.0. Bundesamt fur Sicherheit in der Informationstechnik, 2012
12. Q. Zhang, Z. Li, C. Song, The improvement of digital signature algorithm based on elliptic curve cryptography, in *Proceeding of Artificial Intelligence, Management Science and Electronic Commerce (AIMSEC)*, IEEE, Zhengzhou, China, August 8–10, pp. 1689–1691 (2011)
13. P. Levis, N. Lee, TOSSIM: a simulator for TinyOS networks version 1.0. UC Berkeley, 2003
14. B.L.Titzer, D. Lee, J. Palsberg, Avrora: scalable sensor network simulation with precise timing, in *Proceedings of the 4th International Conference on Information Processing in Sensor Networks (IPSN)*, IEEE, Los Angeles, 2005, pp. 477–482

Chapter 11
Energy Efficient Localization in Wireless Sensor Networks

A.V. Sutagundar, S.S. Shirabur, and V.S. Bennur

Abstract This paper presents an Energy Efficient Localization scheme for Wireless Sensor Networks (EELWSN) based on the received power of beacon signal. The operation of proposed scheme is as follows. (1) Anchor nodes are deployed evenly over the network environment in the predetermined position. (2) Sensor nodes with unknown location are deployed randomly over network environment. (3) Each anchor node broadcasts a beacon signal with location information of the anchor node. (4) Each sensor node should receive more than three beacon signals. (5) Sensor nodes estimate the relative distance between the sensor node and anchor node based on power of each beacon signal received. (6) Sensor node uses trilateration method to compute its position using the relative distances. The performance of the proposed localization scheme is evaluated in terms of performance parameters such as localization error, network lifetime, energy consumption, and cost factor.

11.1 Introduction

In the recent years, the advances in VLSI technology, Micro-Electro-Mechanical system (MEMs) and low power radio technologies have created low power, low cost and multifunctional wireless sensor devices, which can observe and react to changes in physical phenomenon of their environment. WSNs consist of a very large number of small, inexpensive, disposable, robust and low power sensor nodes

---

A.V. Sutagundar (✉) • S.S. Shirabur
Department of Electronics and Communication, Basaveshwar Engineerig College, Bagalkot, Karnataka, India
e-mail: sutagundar@gmail.com; sadashirabur@gmail.com

V.S. Bennur
Department of Electronics and Communication, East Point College of Engineerig for Women, Bengaluru, Karnataka, India
e-mail: vidyabennur@gmail.com

working cooperatively. These sensor nodes are scattered across the geographical area with larger number of nodes to sense over an area so that they can provide greater accuracy[1].

Unfortunately, for a large scale network with hundreds or thousands of sensors the solution is not adding Global Positioning System (GPS) to all nodes in the network and its not feasible for the following reasons [2]. (1) GPS consumes more energy for its operation which inturn reduce the life time of WSN. (2) In large scale WSN, if each sensor node is equipped with the GPS, then the production cost of the network will be very high. (3) Size of sensor nodes are required to be small but the size of the GPS and its antenna increases the sensor node size. (4) In the presence of dense forests, mountains or other obstacles that block the line-of-sight from GPS satellites, GPS cannot be implemented.

Some of the related works of localization technique are as follows: Estimating the geographical position of sensor node is a crucial issue in WSNs to reduce the power consumption, size and cost of wireless device without equipping more number of devices with GPS. The localization from mere connectivity presents a method to determine the location of nodes in the network by using the connectivity information [3]. The work given in [4, 5] describes a localization algorithm based on the angle of arrival of beacons from three or more fixed beacon nodes. In this technique to reduce the hardware complexity it uses direction estimation of received beacons. A multi-hop localization technique for WSNs using acquired Received Signal Strength Indication (RSSI) is presented in [6]. In this technique the RSSI packet is exchanged between the nodes and based on this, a ranging model is constructed, and then it is used by minimum least square algorithm to find location of node.

11.2 Localization

In this section we describe the anchor node deployment and localization scenario in WSNs.

The network environment consists of tiny sensor nodes with heterogeneous sensing capability. In the deployed network, sensor nodes are classified in two types which are sensor node (No prior knowledge of location) and anchor node. The sensor nodes comprises of processor for pre-processing the sensed data, transreceiver for transmission and reception of data and memory. These sensors are deployed randomly over the entire network environment. The anchor nodes comprise of all the components of the sensor node, in addition to that it is consisting of GPS module for location awareness. Anchor nodes are deployed evenly over the entire network environment. After sensor node deployment the localization is performed for the entire network. It is assumed that sensor nodes have same transmission range and capability to reconfigure the transmission power.

11.2.1 Anchor Node Deployment

We consider a $l \times b$ squared area as a network environment for WSNs where ($l = b$). In this squared area 'N_u' identical sensor nodes are deployed randomly and 'N_a' anchor nodes are deployed evenly over the entire network environment. In the proposed scheme anchor nodes are deployed in predetermined position of WSN. The anchor nodes can be deployed based on the Eqs. 11.1 and 11.2.

$$l_x = \frac{l}{z} \times x \quad \text{where } x = 0, 1, 2, 3, \ldots n \tag{11.1}$$

$$b_y = \frac{b}{z} \times y \quad \text{where } y = 0, 1, 2, 3, \ldots m \tag{11.2}$$

Where 'l' is the length and 'b' is breadth of network area, where ($l = b$) and (l_x, b_y) gives the deployment location coordinates of each anchor node by varying the value of 'x' and 'y' in above equation. The anchor nodes can be deployed up till $(l_x, b_y) \leq (l, b)$. The 'z' is a numerical value which should be chosen such that it divides the network area uniformly depending on the number of anchor nodes so that anchor nodes can be deployed evenly over the network. To determine the location information of sensor node, each sensor node should receive beacon signals from minimum three anchor nodes.

To determine the location co-ordinates of sensor nodes the localization is performed. Localization process is described as follows: after the deployment of all sensor nodes, each anchor node broadcasts a beacon signal by flooding (up to two hops). The sensor nodes receive the beacon signals from different anchors. We assume that each sensor node should receive more than three beacon signals from different anchors. Three beacon packets that are received with maximum power are used to compute the distances to their respective anchors from the sensor node.

After estimating the distance to three anchor nodes the trilateration estimation is applied to find the location of sensor nodes. The localization scenario is shown in Fig. 11.1 for one sensor node (U8) and same will be performed by each sensor node.

11.2.1.1 Trilateration

The lateration is a most common method for deriving the position co-ordinates of sensor nodes, where the trilateration uses the distance measurements to three anchor nodes. This estimation considers these distances between the reference location (anchor) and sensor node location as the radii of circles with centres at every respective reference location (anchor location). Thus the sensor location is the intersection of the three sphere surface as shown in Fig. 11.2.

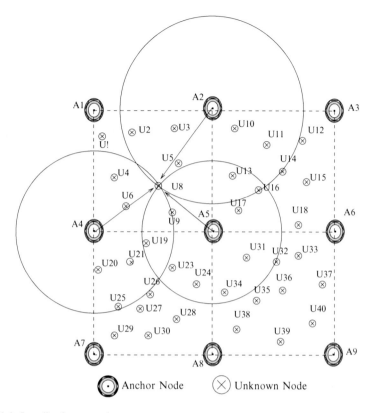

Fig. 11.1 Localization scenario

Fig. 11.2 Trilateration

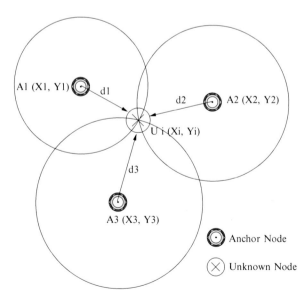

The coordinate (x_i, y_i) of unknown node can be obtained by using estimation method of standard minimum mean square estimation which is given by Eq. 11.3.

$$X = (A^T A)^{-1} A^T \qquad (11.3)$$

11.3 Simulation

To test the performance of the proposed scheme, the scheme is simulated using C programming language. In this section we describe the performance parameters and result analysis.

Some of the performance parameters listed are as follows

- **Localization error**: It is defined as the difference between the estimated and original position of sensor nodes to the total number of unknown nodes.
- **Network lifetime**: It is the total number of rounds taken by the nodes to die in the WSNs.
- **Energy consumption**: It is total amount of energy consumed to determine the location of all the sensor nodes in WSNs.

11.4 Results

This section presents the results obtained during simulation. We compare the results of proposed scheme with existing basic DV-hop localization scheme for WSNs.

Figure 11.3 shows the localization error increases with increase in number of sensor nodes and anchor nodes. The proposed localization scheme provides the reduced localization error compared to basic DV-hop localization algorithm.

Figure 11.4 describes the localization error versus number of sensor nodes with uniform and random placement of anchor nodes. The proposed scheme uses uniform placement of anchor nodes which reduce the localization error as compared to random placement of anchor nodes. When the anchor nodes are deployed randomly, some of the sensor nodes may not get the required number of beacon packets for localization which increases the localization error.

Figure 11.5 presents network lifetime for increase in the number of sensor nodes with and without GPS. As the number of sensor nodes increase, hop distance for communication between the sensor nodes reduce so that network lifetime of sensor nodes increase as seen from simulation results. Communication consumes more power for long distance data transmission and less for shorter distance, so large number of sensor nodes reduces the distance between the sensor nodes.

Figure 11.6 presents the energy consumption for localization in milliJoules with increase in the number of sensor nodes and anchor nodes. As the number of nodes increases the energy consumption decreases. Because of this the communication cost is reduced by smaller hop distances. But the computational cost increases which is negligible as compared to communication cost.

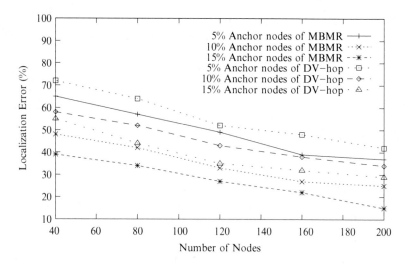

Fig. 11.3 Localization error versus number of sensor nodes

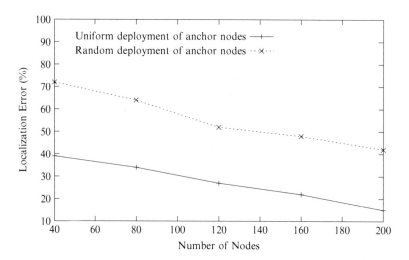

Fig. 11.4 Localization error versus number of sensor nodes with uniform & random placement of anchor nodes

11 Energy Efficient Localization in Wireless Sensor Networks 145

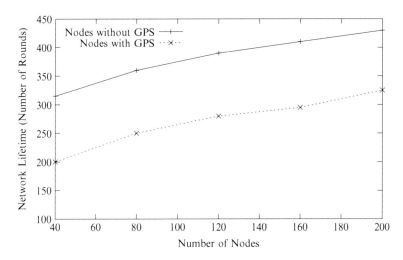

Fig. 11.5 Network lifetime versus number of sensor nodes

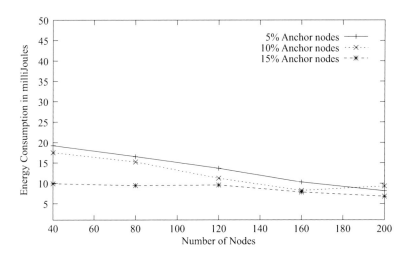

Fig. 11.6 Energy consumption versus number of sensor nodes

11.5 Conclusion

The proposed work presents an energy efficient localization scheme in WSNs by employing trilateration technique to reduce the localization error, energy consumption, size of sensor node and cost of network. The anchor nodes are deployed uniformly to reduce the location error. The location of sensor node is determined by the reception of beacon packets from anchor nodes. The proposed scheme performs better in terms of localization error compared to the basic DV-hop algorithm.

Acknowledgements We are thankful to Basaveshwar Engineering College, Bagalkot, Karnataka, India and TEQIP Phase-2 for the financial assistance.

References

1. I.F. Akyildiz, W. Su, Y. Sankarasubramaniam, E. Cayirci, Wireless sensor networks: a survey. Comput. Netw. **38**(4), 393–422 (2002)
2. A. Pal, Localization algorithms in wireless sensor networks: current approaches and future challenges. Netw. Protoc. Algorithms **2**(1), 45–73 (2010)
3. Y. Shang, W. Ruml, Y. Zhang, M.P. Fromerz, Localization from mere connectivity, in *Proceedings of the 4th ACM International Symposium on Mobile Ad Hoc Networking & Computing*, Annapolis, 2003, pp. 201–212
4. A. Nasipuri, K. Li, Directionality based location discovery scheme for wireless sensor networks, in *Proceedings of the 1st ACM International Workshop on Wireless Sensor Networks and Applications*, New York, 2002, pp. 105–111
5. D. Niculescu, B. Nath, Ad hoc positioning system (APS) using AOA, in *INFOCOM 2003, Twenty-Second Annual Joint Conference of the IEEE Computer and Communications*, IEEE Societies, San Francisco, vol. 3, 2003, pp. 1734–1743
6. C. Alippi, G. Vanini, A RSSI-based and calibrated centralized localization technique for Wireless Sensor Networks, in *Fourth Annual IEEE International Conference on Pervasive Computing and Communications Workshops, PerCom Workshops 2006*, San Diego, 2006, p. 5

Part IV
Network Security, Trust and Privacy

Chapter 12
Data Integrity Verification in Hybrid Cloud Using TTPA

T. Subha and S. Jayashri

Abstract Cloud computing is a growing technology in the field of IT enterprises that provides storage, compute and network resources as a service over the internet. It allows the user to move their application software and large databases to the data centres offered by multiple cloud service providers. Hence trusted enforcement of data and services out sourced on to cloud is a big challenging issue. In this paper we focus on the auditing mechanism in hybrid cloud using trusted Third Party Auditor (TTPA). Hybrid cloud is the one that connects public and private cloud and is useful when the dynamic scalability of service and data migration is needed. This resource expansion is required in the case of storage space limitation at private cloud, further to extend the outsourcing of data in public cloud. We propose a model in which the data is stored in private cloud, moves in to public cloud in case of storage expansion. In order to ensure the integrity and privacy of data, we utilize a Trusted Third Party Auditor (TTPA) to verify the correctness of the data stored in public cloud on behalf of the client. This enables public auditability of data. As a part of our work we carried out a remedy to attain a secure cloud storage services along with high secure data forwarding scheme among the cloud users. The Proposed code based scheme allows the user to verify the CIA (Confidentiality, Integrity and Availability) characteristics of the data with auditing mechanism. It guarantees strong cloud storage correctness, and also simultaneously predicts the misbehaving servers. This scheme is implemented on platform of Amazon web services, a universally accepted cloud vendor and the results are proven effective. This scheme works for all video, image, text files and supports dynamic operations also.

T. Subha (✉)
Department of Information Technology, Anna University, Chennai, India
e-mail: subharajan1979@yahoo.co.in

S. Jayashri
Department of Electronics and Communication Engineering, Adhiparasakthi Engineering College, Chennai, India
e-mail: jayaravi2010@gmail.com

12.1 Introduction

Cloud computing is opening up a new trend in the field of IT enterprises and it offers huge advantages in the IT history such as on-demand self service, ubiquitous n/w access, location independent resource access, rapid elasticity of resource, pricing is based on usage etc [1]. Clients can subscribe for high quality services from IaaS, SaaS and PaaS providers. It is made possible by the increasing network bandwidth and reliable flexible network connections.

The fundamental aspect of this is the data is being centralized or outsourced to the cloud. It brings more benefits from user's perspective by storing data remotely to the cloud. The users are relieved from the burden of storage management, avoidance of capital expenditure on hardware and software, users can able to access the data anywhere, anytime without knowing where their data is actually stored (i.e. location independent access of data) [3]. Public cloud is hosted, operated and managed by third party vendors from one or more data centres. All the services are offered by public cloud can be accessed via the internet through web application or web services [4].

There are three types of deployment models (private, public and hybrid) in cloud. Industries/organizations prefer to use the private cloud for the deployment of their data. The reason is they can have complete control on their data and they can assign role based access control to the employees in their organization. The incorporation of security controls and measures are mandatory in private cloud since it is maintained by the particular organization/industry. In private cloud the storage infrastructure associated is dedicated to a single organization and is not shared with any organizations. It is solely operated by the individual organization or by a third party.

When the clients transfer their data into the public cloud for the cause of saving the storage space and cost, there are concerns about the reliability, confidentiality and accessibility of data [8]. To minimize loss of control over data various encryption technologies can be used [7]. The clients can encrypt their data before outsourcing into cloud. Many authentication mechanisms are used to identify the authorized users who are allowed to access the data in cloud. Mainly security aspects like data integrity, confidentiality and non repudiation are not seen much in private cloud, since it does not allow the unauthorized users to access the data but the public cloud can be accessible and seen by everyone.

These two models can be combined and taking the advantages of both provides a huge benefit for the IT industry is called hybrid cloud. Hybrid cloud consists of multiple internal and external providers from a particular organization. One of the core design principle in cloud is dynamic scalability [11] i.e. it guarantees cloud storage to handle growing amount of applications data in flexible manner. This can be achieved by integrating multiple private and public cloud services that can effectively provide scalability of service and migration of data. In particular organizations might run non-core applications (or) authorized applications in public cloud while maintaining the core applications and sensitive data in house a private cloud [4].

There are many security key challenges needs to be considered when the organizations transfer their infrastructure and data into cloud [10]. The security problem arises mainly because the user does not have control over the data, a lack of trust on the service provider as the provider's data centre stores the data and the location of data is not known to the clients. There also exist legal and multi tenancy issues [8].

The primary goal of our paper is to construct a secure code based scheme for the cloud data storage. In such scenario the CSP (Cloud Service Provider) may behave unfaithfully towards the users. The reason behind is that the users may not retain a local copy of outsourced data and they do not have control over their data [1, 8, 10]. The CSP may in turn provide malicious information regarding the status of their outsourced data.

Hence our first aim is to provide an application based client side encryption scheme for all kinds of outsourced data to CSP. We also consider the functionality for the bulky data that are already uploaded on cloud servers. Therefore our secondary goal is to provide efficient third party auditing methodology to ensure the availability of the data that are outsourced.

12.2 Related Work

In the present scenario the security and governance in the cloud are not satisfactorily maintained. In most of the cloud computing application the cloud user pre-computes Message Authentication Code (MAC) for the file blocks (F) and it is stored locally [8, 10]. It sends the data file F onto the cloud server. During the Audit phase, the data owner releases the key to the server and requests for the new MAC value for the file block. Then this new MAC is compared with the stored MAC to verify the integrity of the data. But the number of times the secret keys are generated is limited in this scheme.

Jachak [9] stated about the problem of ensuring the integrity and security of data storage in cloud computing. The security is achieved in their scheme by signing the data block before sending the data to the cloud. James [12] designed the solution to achieve the storage correctness to ensure users that their data stored in cloud is indeed correct, appropriate and kept unharmed all the time and is retrievable.

Wang [1, 8, 10] proposed a system that includes flexible distributed storage integrity auditing mechanism. They utilized homomorphic token and erasure coded data. It supports dynamic data operations also.

12.2.1 Issues in the Existing Scheme

The previous works prescribed are based on data integrity verification for public cloud as well as hybrid clouds. They suffer from the following reasons that,

1. The solutions proposed by existing schemes are not properly addressed to satisfy the inherent requirement of bandwidth and time.

2. TPA is trusted/independent one.

   (2.a) The existing scheme assumes that TPA involved is trusted one and behaves independently. The auditor maintains small amount of state information for long storage. The auditor somewhat has intention to learn the content of the customer data and the final result derived.

   (2.b) So there may be a possibility of information leakage from TPA or mutual agreement between TPA and CSP (Cloud Service Provider) – it is even possible to change the audit log by CSP, presenting the audit results to the user and make him to believe that the response produced by TPA is correct. So it leads to several problems like the trustworthiness of TPA, it affects the reputation of cloud service provider etc.,

3. The TPA itself may be interested to know the Meta data and tries to infer the relationship from the stored data of clients.

12.3 System Model

Here we propose a hybrid cloud hosting scheme that supports both service scalability and also migration of data from private cloud to public cloud in case the storage requirement is high. We consider the existing multiple cloud service providers to store and maintain the client's data cooperatively. Our main aim of this paper is to maintain the integrity and availability of the data stored in public cloud. It uses the Trusted Third Party Auditing (TTPA) mechanism to verify the data integrity. The proposed architecture is shown in Fig. 12.1.

The following are the identified entities in a system.

Clients: Individual users/Organizations have large amounts of data to outsource into cloud.

Trusted Third Party Auditor: It is a trusted, independent one. It verifies the integrity of the data stored in cloud on behalf of the clients. Whenever the audit request comes from client, the TPA sends the challenge request to the CSP. It then compares the challenge response from the CSP so as to ensure the data integrity and availability.

Cloud Storage Service Provider: CSP has enormous amounts of storage space to store the client's data.

12.3.1 Architecture of Secure Cloud Storage Model

As part of our approach we conducted an experiment for securing the data stored in cloud. This model includes data owner, third party auditor (TPA) other sub-users, cloud vendors. The functionalities of the data owner includes outsourcing the data

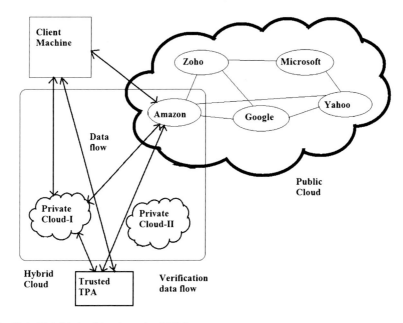

Fig. 12.1 Hybrid cloud hosting using TTPA

files, issuing access credentials to the other user, encrypted upload and data audit delegation. The integrity of the data is verified by TPA after receiving the request from user. It performs public data auditing against the files in the cloud server. The model is shown in Fig. 12.2.

12.3.2 Construction of Our System

Here we describe our approaches for the cloud data storage service scalability and data migration with the above mentioned research goals. Our proposed scheme is implemented on real time architecture by utilizing the Amazon web Services EC2 [2] and simple storage service (S3). It consists of the following important phases:

12.3.2.1 Cloud Processing Functionalities

Authentication

Only authenticated users are allowed to access the portion of the website. The login details of the registered users are stored in RDS provided by Amazon. The users are granted access after verifying these details in RDS during every login attempt.

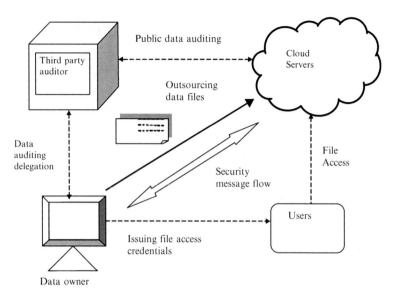

Fig. 12.2 Cloud storage data model

Outsourcing of Data

Initially the File is divided into blocks and the corresponding signatures are added to it. Then the file is outsourced onto cloud. The Meta data is stored with the auditor. The auditor (i.e. verifier) sends the challenge request to the cloud server for integrity verification. The server (i.e. Prover) responds with the proof. The auditor compares both the proof and the original data, Outputs TRUE if it is successfully verified. Otherwise outputs FALSE.

Initialization (or) setup phase – Initially the pre-processing of file F to be outsourced is being carried out in this phase at the data owner's side. It consists of two modules.

Key Generation Algorithm

This is performed at the client's side. The pair of keys called public and private keys is generated using RSA algorithm. Public key is denoted as p_k and the private key is denoted as S_K.

In order to provide security at the user's end, the file F is encrypted using a secret key S_K. It then serves as the way to ensure the confidentiality and integrity of the file F.

Signature (or) Tag Generation Algorithm

This is also performed at the client's side. It generates signature based on the secret key. It uses Merkle Hash Tree authentication structure [1]. It proves the set of elements are not altered and undamaged. It can be constructed as a binary tree where the leaf nodes are the generated hash values of the data values.

First the file F is divided into multiple blocks of equal sizes.

$$F = (m1, m2, m3, \ldots mn) \tag{12.1}$$

Instead of calculating MAC values, this algorithm generates signatures. The advantage is that it avoids the problem of key exhaustion (i.e., If MACs are used to generate data values, the secret keys needs to be stored locally [1]. And it is impractical to calculate new MACs every time due to the communication overhead).

Then signature generation algorithm creates a signature of each block m_i using a secret key and a chosen random value.

Signature set S for all blocks;

$$\mathbf{S}\ (\mathbf{S_i})\ \mathbf{on}\ (\mathbf{m_i})\ \mathbf{for}\ \mathbf{I} = \mathbf{1, 2, n} \tag{12.2}$$

Set of signatures created for a particular file F is the set of ordered hash indexes. It calculates the root R of a MHT, signed using secret key. They are stored in the leaves of a MHT tree. Each file is assigned a file tag say T.

$$\mathbf{T} = \mathbf{name}\,\|\mathbf{nonce}\|\,\mathbf{random\ value}\ \|\ \mathbf{signature\ SK(H(r))} \tag{12.3}$$

The file name, random value, nonce, and signature of these are generated. They are stored at the service provider's side.

Data Auditing Delegation

Many industries and organizations outsource large no of files on to cloud. The process of data integrity and availability verification may be delegated to a third party auditor. This helps the industry to relieve from the burden of checking file integrity frequently, saves time and cost, reduces the resources that are needed to perform frequent integrity check. The TPA is responsible for auditing the clients data stored on the cloud storage. This auditing process should not bring any new vulnerability in terms of data privacy. The auditor should not learn the contents of the user data during the auditing process and by storing the logs.

We introduce a new certificate generation to prove the trustworthiness of auditor. This enables the public auditability of data [6]. A client sends the auditing request to the TPA on a particular file. The validation process is being carried out by TPA. It simultaneously tries to identify the misbehaving server also. In our scheme the users choose the auditor and send the file to TPA to perform the audit. It aims to support the dynamic audit service facility.

Certificate Generation Phase: This phase generates certificates at the client's side, if the data owner wants to check the integrity of its data. This task is delegated to third party auditor. To prove the trust worthiness of TPA, client signs a certificate and sends it to the server and TPA.

$$C = S_{ecret\ key}(name\ ||identity||\ nonce\ TPA) \quad (12.4)$$

Whenever the client wants TPA to perform the audit, the certificate is created and issued every time. This is on demand dynamic auditing scheme. If the TTPA wants to audit the data, TPA sends the certificate C to the server. It verifies and checks the authenticity of TTPA and then allows it to further audit the data integrity.

12.3.2.2 Cloud Auditing Functionalities

Audit Phase

This phase consists of sending challenge request to the server and verifying the challenge response.

Challenge Request: TTPA sends a challenge request on the randomly selected blocks to the cloud storage provider. It also specifies the positions of blocks to be checked.

It takes input as File F; Signature set S, and Challenge. i.e. upon receiving the challenge request the server computes data blocks and corresponding signature blocks for the chosen random subset element. It then returns the data integrity proof as output.

$$\text{Proof} = \text{chalreq}\ (F, S, \text{challenge}) \quad (12.5)$$

Integrity Verification: This algorithm retrieves the signature encrypted using public key p_k. It also calculates the root and authenticates by checking the signatures.

$$\text{Signature S} = \text{chalres}\ (\text{datablock}_i, \text{randomsubset}_c, \text{root}) \quad (12.6)$$

The TTPA compares the response from the provider with the Meta data stored with it. If the test fails it returns the result FALSE. Otherwise ACCEPT.

Audit Report Generation

Finally this module collects the details about the results of audits. It then presents the results and suggestions, the outcome of the audit to the client.

Clients are allowed to access and update their data for various applications dynamically. Authorized applications that allows only authorized users to access and manipulate the data. Our proposed solution aims to provide storage services accountability through independent trusted third party auditing (TTPA). The report is generated to the user's mail id. This is done with the help of the Simple Mail Service provided by Amazon.

12.4 Results and Discussions

Our Experiment is conducted on a real time system architecture implementing the concept in the Amazon web service cloud vendor (AWS EC2). The web application of cloud storage and cloud audit scheme is deployed using the apache tomcat server on the virtual machine. The virtual machines (VM) are provided by AWS as IAAS feature.

This test hosts Linux EC2 instance (1.7 GiB memory, 1 EC2 Compute Unit (1 virtual core with 1 EC2 Compute Unit) 160 GB instance storage, 32-bit or 64-bit platform) for deploying the application. In the given Fig. 12.3a, b the experiments are carried in the hybrid cloud environment and the results are shown [13].

We are implementing in the public cloud and we are conducting more analysis to improve the results than that given in these charts. We are being performing the experiments to improve the computation time overheads.

Snapshots

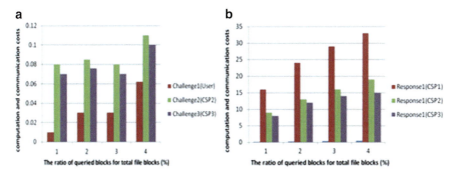

Fig. 12.3 Performance analysis in hybrid cloud environment

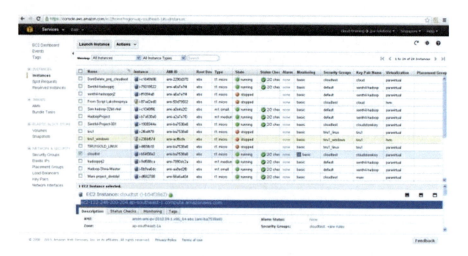

AWS EC2 console with the project running instance

12.5 Conclusion

In this paper we proposed a novel auditing mechanism. The proposed scheme is based on the trusted third party auditor to check data integrity verification for the outsourced data in hybrid cloud. It enhances the trustworthiness of the third party auditor. The auditing result provided by TPA helps to prove the quality of the storage service provider (CSP). CSP in turn can improve the quality of the service and storage to maintain reputation in the competing business market. This scheme proves to be efficient and useful.

References

1. Q. Wang, C. Wang, K. Ren, W. Lou, J. Li, Enabling public auditability and data dynamics for storage security in cloud computing. IEEE Trans. Parallel Distrib. Syst. **22**(5), 847–859 (2011)
2. Amazon.com, Amazon Web Services (AWS) (2009), http://aws.amazon.com
3. Q. Wang, C. Wang, K. Ren, W. Lou, Ensuring data storage security in cloud computing, in *Proceedings of 17th International Workshop Quality of Service (IWQoS'09)*, 2009, IEEE
4. D. Kavitha, G. Divya Zion, Remote sensing data as service in hybrid clouds: security challenges and trusted third party auditing mechanisms. Int. J. Adv. Res. Computer Communication Eng. **1**(7), 481–486 (2012)
5. M. Arrington, Gmail disaster: reports of mass email deletions (2006), http://www.techcrunch.com/2006/12/28/gmail-disaster-reports of-mass-email-deletions
6. G. Ateniese, R. Burns, R. Curtmola, J. Herring, L. Kissner, Z. Peterson, D. Song, Provable data possession at untrusted stores, in *Proceedings of the 14th ACM Conference Computer and Communications Security (CCS'07)*, 2007, pp. 598–609. ACM Newyork

7. M.A. Shah, M. Baker, J.C. Mogul, R. Swaminathan, Auditing to keep online storage services honest, in *Proceedings of the 11th USENIX Workshop Hot Topics in Operating Systems (HotOS '07)*, 2007, pp. 1–6
8. Q. Wang, C. Wang, J. Li, K. Ren, W. Lou, Enabling public verifiability and data dynamics for storage security in cloud computing, in *Proceedings of the 14th European Conference Research in Computer Security (ESORICS '09)*, 2009, pp. 355–370. Springer-Verlag Berlin Heidelberg
9. K.B. Jachak, S.K. Korde, P.P. Ghorpade, G.J. Gagare, Homomorphic authentication with random masking technique ensuring privacy & security in cloud computing BIOINFO security informatics, ISSN: 2249-9423, **2**(2) (2012)
10. K. Ren, C. Wang, Q. Wang, Security challenges for the public cloud. IEEE Internet Comput. **16**(1), 69–73 (2012)
11. M.A. Shah, R. Swaminathan, M. Baker, *Privacy preserving audit & extraction of digital contents*, HP Labs, Technical Report No. HPL-2008-32
12. K. Kajendran, J. Jeyaseelan, J. Jakkulin Joshi, An approach for secured data storage using cloud computing (2011) International Journal of Computer Trends and Technology, ISSN: 2231-2803, pp. 91–96
13. Y. Zhu, S. Wang et al., Secure collaborative integrity verification for hybrid cloud environments. Int. J. Cooper. Info. Syst. **21**(3), 165–197 (2012). World Scientific Publishing Company. doi:10.1142/S0218843012410018

Chapter 13
A Comparative Study of Data Perturbation Using Fuzzy Logic to Preserve Privacy

Thanveer Jahan, G. Narasimha, and C.V. Guru Rao

Abstract The latest advances in the field of information technology have increased enormous growth in the data collection in this era. Individual's data are shared for business or legal reasons, containing sensitive information. Sharing data is a mutual benefit for business growth. The need to preserve privacy has become a challenging problem in privacy preserving data mining. In this paper we deal with a data analysis system having sensitive information. Exposing the information of an individual leads to security threats and could be harmful. The confidential attributes are perturbed or distorted using fuzzy logic. Fuzzy logic is used to protect individual's data to hide details of data in public. Data is owned by an authorized user, and applies distortion. The Authorized user having original dataset distorts numeric data using S-fuzzy membership function. This distorted data is published to the analyst, hiding the sensitive information present in the original data. The analysts perform data mining techniques on the distorted dataset. Accuracy is measured using classification and clustering techniques generated on distorted data is relative to the original, thus privacy is achieved. Comparison of various classifiers is generated on the original and distorted datasets.

13.1 Introduction

In the process of data publishing large volumes of personal data are collected. The increase of technology and global networking database sharing has become a common phenomenon. It can be of criminal records, credit records or a hospital

---

T. Jahan (✉) • G. Narasimha
Department of Computer Science and Engineering, JNTU, Hyderabad, AP, India
e-mail: thanvijahan@gmail.com; narsimha06@gmail.com

C.V. G. Rao
Department of Computer Science and Engineering, S.R Engg, Warangal, AP, India
e-mail: guru_cv_rao@hotmail.com

releasing patient's record. Data is sensitive to privacy issues. Defense applications, financial transactions, healthcare records and network communication traffic [1]. The researchers or data analysts use these data to analyze by data mining techniques. Data mining is the process of gathering and collecting data to extract information. Analyzing such raw data can cause threat to privacy. Data containing sensitive or confidential information is protected using privacy preserving data mining. Many approaches have been employed in preserving privacy Randomization, Anonymization and secure multiparty computation. Randomization method consists of data perturbation or data modification which perturbs the confidential attributes. Classes of methods are proposed for privacy protection in data processing that is used in analysis system. Data perturbation methods are used to modify data or add noise to data [2], data mining techniques have proved that original and perturbed data are relatively same and accuracy is measured by different classifiers. The dimensionality of the matrix is reduced by transforming original dimension of data. Wang et al. [3] suggested significance of feature selection for analysis purpose and suggested that performing SSVD and feature selection is a better approach for classification purpose, while discarding features having small distorted values. Various methods are adopted for preserving privacy such as data swapping [4, 5] the attributes are interchanged with a higher probability. In Aggregation [6] the row is represented as group of values. The Fourier and signal Transformation [7, 8] methods are fast improving time complexity. In data Anonymization different approaches such as generalization and suppression methods are used, while k-anonymity protects identity disclosure but not attribute. In secure multi party computation (SMC) [9] data is encrypted using protocols such as secure sum, secure union and secure without revealing private data to the data miners.

B. Karthikeyan et al. [10] used fuzzy membership function on original data, proved efficient increase and decreased the number of passes to perform clustering. In this paper we extend our work on an application where the information is imprecise and fuzzy logic provides better solution [5]. The individual information is preserved revealing details in public using fuzzy reasoning. The confidential attributes are modified using s-based horizontally distributed data by performing union of all individual entities. The distorted data is analyzed using data mining techniques such as classification. Numbers of methods are used to preserve privacy which increases complexity and processing time. An optimum solution is achieved in this paper using fuzzy based approach. The rest of the paper is organized as follows: Sect. 13.1 is the literature survey, Sect. 13.2 is the background work on privacy preserving data mining. Section 13.3 describes fuzzy based approach used in privacy preserving. Sections 13.4 and 13.5 are the classification and Clustering used on datasets. Section 13.6 describes about the proposed method, experimental results and the comparison between the classifiers and clustering on original and distorted datasets, and finally Sect. 13.7 sums up with conclusion of the work proposed and future work.

13.2 Background and Related Work

13.2.1 Privacy Preserving Data Mining

The main aim of preserving privacy is hiding sensitive data, while it is been published. The raising concern of privacy had led disclosure of information. Data can reside at a single organization or in different places i.e. distributed data. In such scenarios relevant algorithms are used to protect data in privacy preserving data mining (PPDM). Many approaches are adopted to solve these issues, developing algorithms to modify the original data in such a way that data and knowledge remain private even after mining process [11]. Techniques include data perturbation, blocking feature values, swapping tuples etc. PPDM scheme should able to maximize the degree of data modification to retain the maximum data utility.

13.2.2 Analysis System and Perturbation

A Data analysis model is shown in Fig. 13.1 consists of two parts an authorized user and data analyst [12]. The authorized user owns an original data and manipulates the data. Data is represented in tabular form having rows and columns. The original data has sensitive information and should be disclosed for privacy. Authorized user manipulates original data into perturbed data. The perturbed data is called as a fuzzy data. Fuzzy data hides the sensitive information of an original data. During data publishing user gives fuzzy data to data analyst. Data analyst collects fuzzy data to perform data mining techniques. In this way data is protected by an authorized user distorting the actual values by fuzzy values. Data mining techniques used by analyst are classification, clustering.

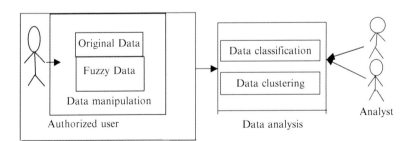

Fig. 13.1 Data analysis system

13.3 Fuzzy Based Approach

Fuzzy sets are the extension of generic set theory, it is introduced in Ref. [13] has a different approach to preserve privacy. The main characteristics of fuzzy sets contrasting with crisp set, is the progressive transition from one set to another. The natural characteristic of fuzzy logic provide automatic mechanism to deal with imprecision and uncertainty, which are inherent to real world knowledge. The assessment of data set can be done using fuzzy membership in fuzzy sets [14]. A fuzzy set is a pair (A, μ$_A$) where A is a set and μ$_A$: A → [0, 1]. For all x ϵ A, μ$_A$(x) is called the grade of membership of x. Each linguistic term can be represented as a fuzzy set having its own membership function [15]. An S-shaped fuzzy membership function is given as:

$$f(x; a, b) = \begin{cases} 0, & x \leq a \\ 2\left(\frac{x-a}{b-a}\right)^2, & a \leq x \leq \frac{a+b}{2} \\ 1 - 2\left(\frac{x-b}{b-a}\right)^2, & \frac{a+b}{2} \leq x \leq b \\ 1, & x \geq b \end{cases}$$

Where x− is the value of the sensitive attribute, a & b are the minimum and maximum value of the sensitive attribute in the original data set.

13.4 Classification

Data mining utilities are used to assess an original dataset and dataset after perturbation. The analyst performs data mining techniques such as classification, clustering on distorted data. In this paper we used various classifiers such as SVM, ID3 and C4.5 on original data and perturbed data. The accuracy results have found the best classifier among them. The above graph proves that SVM gives promising accuracy results than ID3 and C4.5. The data before and after perturbation is relatively same and is proved by mining utility. Classification is a process of finding a set of models that describe and distinguish data classes and concepts. The purpose of being able to use model is to predict class, where label is unknown. Classification is a two step process shown in Fig. 13.2. (1) Build classification model using training data. Every object of the data must be pre-classified i.e. its class label must be known. (2) The model generated in the preceding step is tested by assigning class labels to data objects in a test dataset. The test data may be different from the training data. Every element of the test data is also pre-classified in advance. The accuracy of the classification model is determined by comparing true class labels in the testing set with those assigned by model.

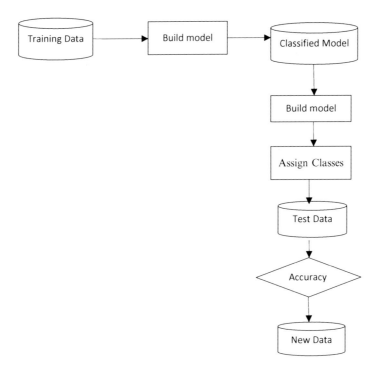

Fig. 13.2 Classification process

13.5 K-Means Clustering

Clustering is a well-known problem in statistics and engineering, namely, how to arrange a set of vectors (measurements) into a number of groups (clusters). Clustering is an important area of application for a variety of fields including data mining, statistical data analysis and vector quantization [16]. The problem has been formulated in various ways in the machine learning, pattern recognition optimization and statistics literature. The fundamental clustering problem is that of grouping together (clustering) data items that are similar to each other. Given a set of data items, clustering algorithms group similar items together. Clustering has many applications, such as customer behavior analysis, targeted marketing, forensics, and bioinformatics.

13.6 Experimental Results

In this paper we have used a real world datasets Fertility, Hepatitis and Iris datasets downloaded from UCI machine learning Repository having details of patients of hepatitis. These Datasets have the sensitive attribute an age of the patients.

The sensitive attribute is transformed into a distorted data. This distorted data is published protecting privacy of an individual. The original dataset has sensitive information about patient is perturbed with S-based fuzzy membership function. In our experiments we have used Tanagra data mining tool for classification, k-means clustering is implemented in JAVA and performance is checked using MATLAB package.

13.6.1 Proposed Method

Step 1: An authorized user owns an original dataset (D).
Step 2: A original dataset having sensitive attributes is perturbed using S-based fuzzy membership function (fuzzy data) ($\overline{D}$).
Step 3: The fuzzy data ($\overline{D}$) is published by a user to an analyst for analysis.
Step 4: Analyst receives the fuzzy data and performs mining techniques.

The different classifiers used are SVM, ID3 and C4.5. The data before and after perturbation is relatively same, proved by mining utility. "Accuracy of a classifiers is simply, a ratio of ((no. of correctly classified examples)/(total no. of examples)) *100)".

Technically it can be defined

$$accuracy = \frac{TP+TN}{(TP+FN)+(FP+TN)}$$

An experiment measuring accuracy of classifiers based on True Positives (TP), False Positives (FP) as per the above equation is tabulated in Table 13.1. The tabular form indicates the accuracy of original and perturbed dataset on classifiers SVM, ID3 and C4.5. The results tabulated indicate that classification performed on original data and perturbed data are equivalent. The Accuracy of classifiers and k-means clustering is shown in Figs. 13.3, 13.4 and 13.5. The results indicate that classification and clustering performed on original data and perturbed data are relatively equivalent. We have found that by using fuzzy approach, the processing time of data is considerably reduced when compared to the other methods that were used before.

We have found that by using fuzzy approach, the processing time of data is considerably reduced when compared to the other methods that were used before.

Table 13.1 Classification of datasets

Data	Iris dataset							Hepatitis							Fertility data set						
CLASS	ID3		SVM		C4.5			ID3		SVM		C4.5			ID3		SVM		C4.5	
	TP	FP	TP	FP	TP	FP		TP	FP	TP	FP	TP	FP		TP	FP	TP	FP	TP	FP
ORIG	0.98	0.02	0.97	0.03	0.99	0.01		0.92	0.07	0.92	0.7	0.89	0.10		0.88	0.12	0.88	0.12	0.87	0.07
DIST	0.98	0.02	0.97	0.03	0.99	0.01		0.92	0.07	0.92	0.07	0.89	0.10		0.87	0.13	0.86	0.14	0.87	0.07

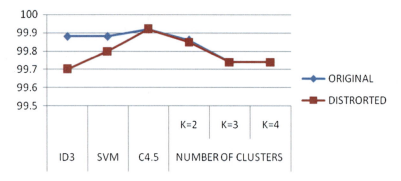

Fig. 13.3 Fertility dataset

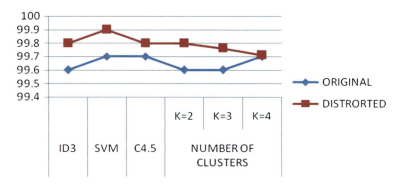

Fig. 13.4 Iris dataset

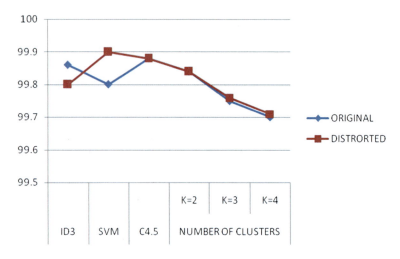

Fig. 13.5 Hepatitis dataset

13.7 Conclusion and Future Scope

The paper presents a fuzzy based approach to transform a original data to fuzzy data i.e. perturbed data. The method is proved an efficient maintaining privacy while it is published. The analyst is unknown by original values, hence preserving privacy of sensitive information owned by an authorized user. The results from our experiments shows that classification performed on original and perturbed data are relatively same. Fuzzy approach S-based membership function used has increased processing time of the algorithm used. In future we would like to extend our work other fuzzy membership functions such as triangular and use other classification and clustering data mining utilities for the proposed algorithm used in this paper.

References

1. V. Estvill–Castro, L. Brankovic, D.L. Dowe, Privacy in data mining. Australian Computer Society NSW Branch. Available at www.acs.org.au/nsw/articles/199082.html
2. Thanveer, G. Narasimha, C.V. GuruRao, Data Perturbation and Feature Selection in Preserving Privacy, in *Proceeding of the 2012 Ninth International conference in Wireless and Optical Communication Networks(WOCN)*, (Indore). IEEE catalog number: CFP12604-CDR, ISBN: 978-1-4673-1989-8/12
3. Pengpeng Lin, Jun Zhang, Ingrid St. Omer, Huanjing Wang, Jie Wang, in *A Comparative Study on Data Perturbation with Feature Selection*. Proceeding of the international multiconference of engineers and computer scientists 2011, vol. 1 (IMECS, Hongkong, 2011), 16–18 Mar 2011
4. S.E. Fienberg, J. McIntyre, Data swapping: variations on a theme by Dalenius and Reiss. J. Off. Stat. **21**, 309–323 (2005)
5. K. Muralidhar, R. Sarathy, Data shuffling a new masking approach for numerical data. Manage. Sci. **52**, 658–670 (2006)
6. Y. Li, S. Zhu, L. Wang, S. Jajodia, Privacy enhanced micro aggregation method, in *Proceedings of 2nd International Symposium on Foundations of Information and Knowledge Systems*, 2002, pp. 148–159
7. Shuting Xu, Shuhua Lai, Fast Fourier Transform based data perturbation method for privacy protection, in *Proceedings of IEEE Conference on Intelligence and Security Informatics*, New Brunswick New Jersey, May 2007
8. S. Mukharjee, Zhiyuan Chen, A. Gangopadhyay, A privacy preserving technique for Euclidean distance-based mining algorithms using Fourier-related transforms, VLDB J. **15**, 293–315 (2006)
9. Pinkas, Cryptographic techniques for privacy- preserving data mining, ACM SIGKDD Explorations, **4**(2), 12–19 (2002)
10. B. Karthikeyan, G. Manikandan, V. Vaithiyanathan, A fuzzy based approach for privacy preserving clustering. J. Theor. Appl. Inf. Technol. **32**(2), 118–122 (2011)
11. R. Agrawal, R. Srikant, Privacy–preserving data mining, in *Proceedings of the 2000 ACM SIGMOD International Conference on Management of Data*, San Diego, 2003, pp. 86–97
12. S. Xu, J. Zhang, D. Han, J. Wang, Data distortion for privacy protection in a terrorist analysis system, in *Proceedings of the 2005 I.E. International Conference on Intelligence and Security Informatics*, 2005, pp. 459–464

13. V. Vallikumari, S. Srinivasa Rao, KVSVN. Raju, KV. Ramana, BVS. Avadhani, Fuzzy based approach for privacy preserving publication of data. Int. J. Comput. Sci. Netw. Secur. **8**(1), (2008)
14. L. Zadeh, Fuzzy sets. Inf. Control. **8**, 338–353 (1965)
15. J. Timothy, *Ross, Fuzzy Logic with Engineering Applications* (McGraw Hill, New York/Singapore, 1997)
16. T. Jahan, G. Narsimha, C.V Guru Rao, Privacy preserving clustering on distorted data in International Organization of Scientific Research. J. Comput. Eng. ISSN: 2278–0661, ISBN: 2278–8727 **5**(2), 25–29 (2012)

Chapter 14
Vector Quantization in Language Independent Speaker Identification Using Mel-Frequency Cepstrum Co-efficient

D. Ambika and V. Radha

Abstract Speaker recognition is a process of recognizing a person based on their unique voice signals and it is a topic of great importance in areas of intelligent and security. Considerable research and development has been carried out to extract speaker specific features and to develop features matching techniques. The goal of this paper is to perform text-independent speaker identification. These models rely on Mel Frequency Cepstral Coefficients (MFCC) for extraction of speaker specific features and for speaker modelling Vector Quantization (VQ) is used due to high accuracy and simplicity. The proposed system efficiency was analyzed by using 20 filter banks for extracting features. The performance was evaluated using MATLAB against different speakers in different languages such as Tamil, Malayalam, Hindi, Telugu and English with duration of 2, 3 and 4 s. Experimental result shows that 4 s duration of speech regardless of language is able to produce 98 %, 99 % and 97 % of identification when compared to 2 and 3 s. The system efficiency may further be improved using other speaker modelling techniques like Neural Network, Hidden Markov Model and Gaussian Mixture Model.

14.1 Introduction

Speech has been considered to be a most important component for human to convey information to one another. The speech signal conveys several types of information such as the emotion, gender, accent, the language being spoken and generally the identity of the speaker. As a result, the acoustic signal of human speech characterizes not only what is being said but also it represents individual characteristics of

---

D. Ambika (✉) • V. Radha
Department of Computer Science, Avinashilingam Institute for Home Science and Higher Education for Women, Coimbatore, India
e-mail: ambikaphdscholar@gmail.com

the speaker such as the individual pitch, vocal tract resonances as well as speaking styles and durations. The speech perception conveys information about the environment in which the speech was produced and transmitted. The speech recognition recognizes the word spoken in speech, and the speaker recognition system is used to extract, characterize and recognize the information in the speech signal which conveys speaker identity. By using the speaker-specific information, speaker recognition automatically recognizes the speaker included in speech waves to verify identities being claimed by people accessing systems; that is, it allows access control of various services by voice [1–3]. It is a fundamental part of oral communication between humans. No two individuals sound the same because of their individual vocal tract shapes and larynx sizes. In addition to that, every individual has his own distinctive manner of speaking, including the use of a particular tone of voice, rhythm, modulation style, pronunciation model, choice of expressions and so on.

Sometimes, there may be chances for the automatic speaker recognition system to make decision errors. There are many sources of variation which may contribute to cause errors, in which some of them are basic physical attributes, language, accent, characteristics of speaking style, and changes in emotional state or health. The factors which affect the speaker recognition system performance are [4]:

- Speech quality: depends on the types of microphones used, ambient noise levels, types of noise, etc.
- Speech modality: text-dependent or text–independent
- Speech duration: the time required for training and testing data, temporal division of training and testing data.
- Speaker population: number and similarity of speakers.

There are many practical difficulties available for speaker recognition process, in which some of them are: high performance under requirements for Robustness and flexibility, initial training, adaptation, decision strategy, human behaviour and performance, etc. Yet, one of the main challenges remains the human factor. There exist two types of speaker recognition such as: the naive speaker recognition and the technical (automatic or semi-automatic) speaker recognition. In naïve method, the recognition is performed by untrained observers (a human "expert"). The decision is based on what is heard and no special techniques are involved. Whereas in the second method a great deal of work on the comparison of recordings is required.

14.2 Types of Speaker Recognition

Speaker recognition is the process of automatically recognizing the speaker voice according to the basis of individual characteristics information in the voice waves. It can be defined as any activity whereby a speech sample is attributed to a person on the basis of its phonetic, acoustic or perceptual properties. The speaker

14 Vector Quantization in Language Independent Speaker Identification... 173

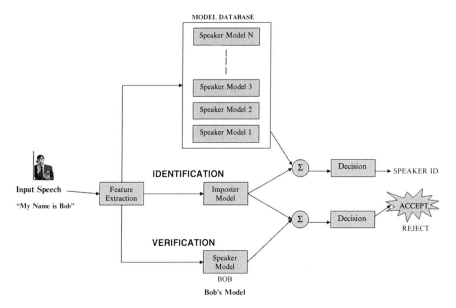

Fig. 14.1 Basic structure of speaker recognition system

recognition can be classified into two main areas such as identification and verification. Speaker identification is a 1: N match where the voice is compared against N templates. In a speaker verification 1:1 match is done where one speaker's voice is matched to one template. It is also known as speaker authentication or detection and it is the task of determining whether a person who he/she claims to be (a yes/no decision). The basic structure of speaker recognition system is given in Fig. 14.1.

The speaker identification process may be text dependent (restrained) or text independent (unrestrained). In text dependent speaker identification system, the speaker is asked to utter a specific phrase (password) which is the same for enrollment and for verification, In text independent speaker identification system, the recognition is based on identifying speaker irrespective of any utterance.

There are two modes of operation such as open set and closed set mode. In closed set systems, the speaker is known a priori to be a member of a set of finite speakers. This type of identification can be considered as a multiple-class classification problem. In open set mode, the speaker can be an outsider and not necessary he/she should be from the set of already defined speakers. The speakers that are not present in the set of known voices can be called as impostors in open set mode. This task can be used in forensic applications, e.g., speech evidence can be used to recognize the perpetrator's identity among several known suspects. This paper takes speaker identification into consideration, which consists of mapping a speech signal from an unknown speaker to a database of known speakers, i.e. the system has been trained with a number of speakers in which it has to recognize.

14.3 Speaker Recognition Applications

Speaker recognition technologies have wide application areas and it is quietly varied and continually growing. The applications of voice verification includes government, healthcare, call centers, electronic commerce, financial services, customer authentication for service calls, for house arrest and probation-related authentication. Mostly it is used for voice biometric systems. The typical applications of speaker recognition system are given in Table 14.1.

14.4 Recognition System

The general approach to Automatic speaker identification consists of various steps such as digital speech data acquisition, feature extraction and classification.

14.4.1 Data Collection

Voice recording can be performed either using a local dedicated system or remotely (e.g. telephone) but it depends upon the application. The acoustic patterns of speech can be pictured as loudness or frequency vs. time. Speaker recognition systems analyze the frequency as well as attributes such as dynamics, pitch, duration and loudness of the signal.

14.4.2 Feature Extraction

The feature extraction is a data reduction process which tries to capture the vital characteristics of the speaker with a small data rate. A sample extracted feature using MFCC is given in Fig. 14.4. It is an important component in speech and speaker recognition because the accuracy of recognition mainly depends on the

Table 14.1 Typical applications of speaker recognition systems

Areas	Specific applications
Authentication	Remote identification and verification, mobile banking, ATM transaction, access control
Information security	Personal devices logon, desktop logon, application security, database security, medical records, security control for confidential information
Law enforcement	Forensic investigation, surveillance applications
Interactive voice Response	Banking over a telephone network, information and reservation Services, telephone shopping, voice dialing, voice mail

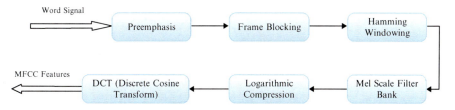

Fig. 14.2 MFCC feature extraction process

features that are extracted. Some of the important properties that feature extraction techniques should satisfy are:

- High noise and distortion robustness
- High disguise and mimicry robustness
- High inter-speaker variation
- Low intra-speaker variation
- Easy to measure
- Maximally internally independent features.

Different techniques available for feature extraction such as Linear Predictive Coding (LPC), Linear Predictive Cepstral Coefficients (LPCC) and Mel Frequency Cepstral Coefficients (MFCC). In this research work the most frequently used MFCC parameters are considered to determine the best feature set. The feature extraction involves various steps such as pre-emphasis, frame blocking, windowing, Mel scale filter bank analysis, logarithmic compression and discrete cosine Transformation (DCT). The overall process of MFCC feature extraction process is shown in Fig. 14.2.

14.4.2.1 Pre-emphasis

Pre-emphasis filter is one of the common filters used in speech enhancement which reduces background noise and resulting in good quality of speech. The original signal before and after pre-emphasis for the sample word is shown in Fig. 14.3. The pre-emphasis filter spectrally flattens the speech signal to improve efficiency of speech waveform. Typical signal pre-emphasis is given by [5]

$$S2(n) = s(n) - a^*s(n-1) \qquad (14.1)$$

Where constant 'a' falls between intervals 0.9–1.0.

14.4.2.2 Frame Blocking

The extracted speech from a window is called a frame; the duration of the sampling is called the frame size and the time between successive frames is called frame

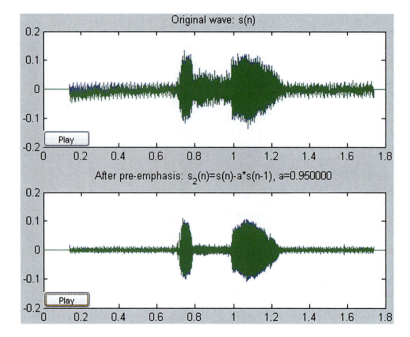

Fig. 14.3 A sample speech waveform before and after pre-emphasis filter

overlap (frame shift). The speech signal can be separated into short fixed length frames. Where the continuous speech signal is also divided into frames, in which each frame consists of N samples overlapping with each other by M samples [6].

14.4.2.3 Windowing

Windowing is carried out after frame segmentation in order to minimize the spectral distortion in which window is used to taper the signal on both ends. So it reduces the side effects caused by signal discontinuity, at the beginning and at the end due to framing. There are mainly three types of windowing functions namely: (i) rectangular, (ii) hamming and (iii) hanning window.

For this work hamming window is used which is multiplied with each frame:

$$w(n) = 0.54 - 0.46 \cos\left(\frac{2n\pi}{N-1}\right) \quad 0 \leq n \leq N-1 \qquad (14.2)$$

where N is the number of samples in each frame.

14.4.2.4 Filter Bank Analysis

A Fast Fourier Transform (FFT) of speech signal is wide and it does not follow a linear scale so magnitude is weighted by the series of filter frequency responses. In fact, the frequency scale of cochlea in the human ear is actually non-linear which is also known as mel scale (a mel is the unit of a pitch). This scale has linear frequency spacing below 1 kHz and a logarithmic spacing above this value. Filter bank analysis is a process of converting time domain speech signals of frame of N samples to frequency domain. The mel-frequency can be calculated using the equation

$$Mel\ (f) = 2595 * log\ 10\ (1 + f/700) \qquad (14.3)$$

14.4.2.5 Logarithmic Compression

The outputs obtained from filter bank analysis are compressed using logarithmic function.

$$X_m\ (ln) = ln\ (X_m) \quad 1 \leq m \leq \qquad (14.4)$$

Where $X_{m(ln)}$ is logarithmically compressed output of mth filter.

14.4.2.6 Discrete Cosine Transformation

The first few coefficients grouped together as a feature vector of a particular speech frame by applying Discrete Cosine Transformation for filter outputs (Fig. 14.4). The kth MFCC coefficient in the range $1 \leq k \leq p$ can be expressed as [7]

$$MFCC_k = \sqrt{2/M} \sum X_{m(ln)} cos\ (\pi k\ (m - 0.5)\ M) \qquad (14.5)$$

Where p is the order of Mel scale spectrum.

14.4.3 Classification

It involves two phases such as speaker modelling and speaker matching.

14.4.3.1 Speaker Modeling Using Vector Quantization

Speaker modeling is a process of mapping vectors to a finite number of regions from a large vector space in that space. Where each region is called a cluster and

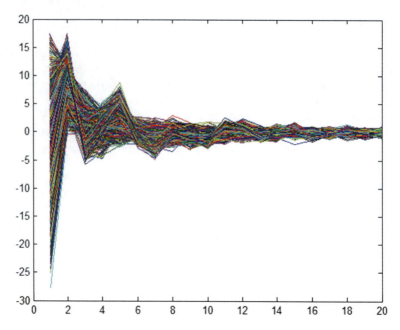

Fig. 14.4 Feature vectors using MFCC

can be represented by its center called a codeword. The codeword is used to serve as the model of the speaker and the collection of all code words can be called as codebook [8]. The number of code vectors is considerably smaller than the training set and by clustering his/her training acoustic vectors, a speaker-specific VQ codebook is generated for each known speaker in a multidimensional space. Each cell defines a small part of the total space and contains a point centered within the cell, the centroid [9]. Hence, a vector quantizer Q of dimension k and size N is a mapping from a vector in the k dimensional space into one of N centroid in the space.

14.4.3.2 K-Means Clustering

This algorithm is used to group data vectors into K groups based on features. The K-means algorithm [6, 10] was developed mainly for the Vector Quantization codebook generation. It characterizes each cluster by the mean of the centroid using vector. By minimizing the sum of squares of distances between the data vectors and the corresponding cluster's centroid, the grouping of data can be done [8].

14.4.3.3 Linde-Buzo-Gray (LBG) Clustering

This [10] algorithm is a finite sequence of steps in which, at every step, a new quantizer, with an average distortion less or equal to the previous one, is produced. There are two phases such as, the initialization of the codebook and its optimization. The optimization starts from the initial codebook, and after some iteration, a final codebook is generated with a distortion subsequent to a local minimum.

14.5 Experimental Results and Discussion

The speech samples were collected using different languages such as English, Tamil, Telugu, Malayalam and Hindi. The amount of speech given for training and testing the speaker identification system is 2, 3 and 4 s, in order to analyze the speech signal hamming window is used. Figure 14.5 shows the acoustic vectors distribution of different samples for three users using MFCC signals. In Fig. 14.6 the acoustic distribution of a user for three different samples using MFCC is shown. In the Fig. 14.7 the code vectors of three different speakers are marked by red, pink, cross blue colours and codebook is generated for three different users. The proposed system efficiency was analyzed by using 20 filter banks for extracting features.

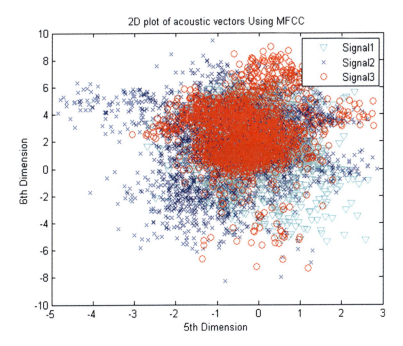

Fig. 14.5 2D acoustic vectors distribution of different users using MFCC

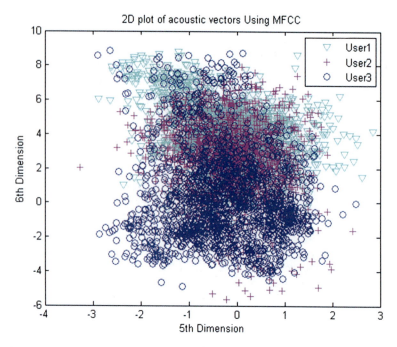

Fig. 14.6 2D acoustic vectors distribution of a user for three different samples using MFCC

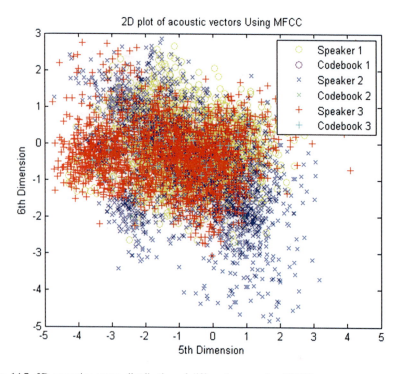

Fig. 14.7 2D acoustic vectors distribution of different users using MFCC

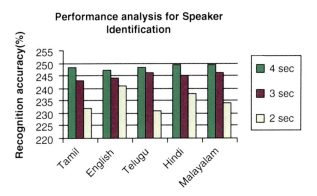

Fig. 14.8 Performance analyses for speaker identification

Table 14.2 Summary of speaker identification rate

Voice sample	Language	Speakers	No. of utterance	Correct identification	Accuracy
4 s	Tamil	25	250	248	99.2 %
3 s		25	250	243	97.2 %
2 s		25	250	232	92.8 %
4 s	English	25	250	247	98.8 %
3 s		25	250	244	97.6 %
2 s		25	250	241	96.4 %
4 s	Telugu	25	250	248	99.2 %
3 s		25	250	246	98.4 %
2 s		25	250	231	92.4 %
4 s	Hindi	25	250	249	99.6 %
3 s		25	250	245	98 %
2 s		25	250	238	95.2 %
4 s	Malayalam	25	250	249	99.6 %
3 s		25	250	246	98.4 %
2 s		25	250	234	93.6 %

The Fig. 14.8 shows the performance of speaker identification in a graphical form and their identification rate is given in Table 14.2. Here the voice samples were the different duration of the speech samples taken from 25 speakers. The utterances were collected in Indian languages such as Tamil, Telugu, English, Hindi and Malayalam. The speech samples were text independent. From each speaker ten samples were collected such as No. of speakers = 25, No. of utterance = 10, so 25 * 10 = 250 utterances were collected for each language as well as for each duration of seconds. From the result it can be analysed that irrespective of language, the identification system is able to produce high accuracy with 4 s when compared to others. This is because when more duration and more training are given, the system will be able to produce high accuracy. Within 2 s the user or the speaker will be able to produce only a limited utterance, for example, one or two utterances. That is why the accuracy reduces.

14.6 Conclusion

Speaker recognition is the process of recognizing people from their unique voices. In this paper feature extraction based on MFCC is used for speaker identification and their performance was evaluated against different speakers in different languages with different durations. The result shows that 4 s duration of speech, regardless of language is able to produce 98 %, 99 % and 97 % of identification when compared to 2, 3 and 4 s. The proposed system efficiency may further be improved by using other speaker modeling techniques like HMM, GMM and Neural Networks.

References

1. S. Furui, Speaker-independent and speaker-adaptive recognition techniques, in *Advances in Speech Signal Processing*, ed. by S. Furui, M.M. Sondhi (Marcel Dekker, New York, 1991), pp. 597–622
2. S. Furui, Recent advances in speaker recognition, in *Proceedings of the First International Conference on Audio- and Video-Based Biometric Person Authentication*, 1997, pp. 237–252
3. S. Furui, *Digital Speech Processing, Synthesis, and Recognition*, 2nd edn. (Marcel Dekker, New York, 2000)
4. D.A. Reynolds, *Automatic Speaker Recognition: Current Approaches and Future Trends* (MIT Lincoln Laboratory, Lexington, 2006)
5. M. Sigmund, *Voice Recognition by Computer* (Tectum Verlag DE, Marburg, 2003)
6. H.S. Jayanna, S.R.M. Prasanna, Analysis, feature extraction, modeling and testing techniques for speaker recognition. IETE Tech. Rev. **26**, 181–190 (2009)
7. A.N. Sigappi, S. Palanivel, Spoken word recognition strategy for Tamil language. IJCSI Int. J. Comput. Sci. Issues, **9**(1), No. 3 (2012). ISSN 1694-0814
8. M.G. Sumithra, K. Thanuskodi, A new speaker recognition system with combined feature extraction techniques. J. Comput. Sci., **7**(4), 459–465 (2011), Science Publications. ISSN 1549-3636
9. Y. Goto, T. Akatsu et al., An investigation on speaker vector-based speaker identification under noisy conditions, in *Proceedings of the International Conference on Audio, Language and Image Processing*, IEEE Xplore, pp. 1430–1435
10. S. Menon, M. Lech, N. Maddage, Speaker verification based on different vector quantization techniques with Gaussian mixture models, in *Proceedings of the 3rd International Conference on Network and System Security*, pp. 403–408

Chapter 15
Improved Technique for Data Confidentiality in Cloud Environment

Pratyush Ranjan, Preeti Mishra, Jaiveer Singh Rawat, Emmanuel S. Pilli, and R.C. Joshi

Abstract Cloud Computing allows anyone to provision virtual hardware, runtime environment and services. Resources can be accessed over the Internet and offered on pay-per-use basis by cloud computing service providers (CSPs). Data storage is one of the services of cloud computing. Data Security is important concern in cloud. Confidentiality, Integrity, and Availability (CIA) are some security dimensions. In this paper, we are providing an improved technique to provide confidentiality of data. The proposed encryption and decryption algorithm achieves better time and storage complexity than the existing ELGamal algorithm. We have implemented this algorithm in Java and run this in different data set available in internet.

15.1 Introduction

A cloud is a type of parallel and distributed system consisting of a collection of interconnected and virtualized computers that are dynamically provisioned and presented as one or more unified computing resources based on Service Level Agreement (SLA) defining quality of service parameters under which the service is delivered [1].

The three main entities of cloud are CSP (Cloud Service Provider), Client/Owner, User. Cloud Service Provider manages the Cloud Storage Server (CSS). It has significant resources and high computing power. Organization/Client entity has large data files to be stored in the cloud and depends on cloud for data maintenance and computation. User is registered with the owner and uses the data of owner stored on the cloud.

IaaS is one of the popular cloud services. Client may have terabytes or more data. Storing such a huge amount of data in personal hard disk or other storage

---

P. Ranjan (✉) • P. Mishra • J.S. Rawat • E.S. Pilli • R.C. Joshi
Department of Computer Science and Engineering, Graphic Era University, Dehradun, India
e-mail: pratyushranjan@live.com; preeti.mish22@gmail.com; jairwt445@gmail.com; emmshu@gmail.com; chancellor.geu@gmail.com

device is very expensive. Instead of storing information in a personal HDD, client may save it to a remote database where internet provides a connection between the user computer and remote database.

Cloud service providers (CSPs) and consumers both gets the benefit of cloud computing as it has got the interesting features like efficient resource allocation, dynamic resource provisioning, on demand access, pay-per-use pricing scheme, energy efficiency, scalability, multi tenancy etc.

A storage area network (SAN) is a dedicated high speed network for block level data access. It carries data between servers (also known as hosts) and storage devices through Fiber Channel switches. SAN enables storage consolidation and allows storage to be shared across multiple servers. SAN provides the physical communication infrastructure and enables secure and robust communication between host and storage devices.

Maintenance cost and operational cost is reduced related to IT software and infrastructure. Hence cloud technology provides a good economical return to the company. The resources are prone to number of attacks like denial of service, session hijacking, identity theft, web application attack, data stealing and data leakage etc.

Alert cloud Security report categorizes the attacks as follows: web application attack 52 %, Brute force attack 30 %, and vulnerability 27 %. In Enterprise data center Malware/Botnet activity 49 %, Brute force 49 %, and web application attack 39 %. Enterprise data centers are more likely to be struck by targeted rather opportunistic attack whereas as the opposite is true in case of CSPs data centers [2].

Personal data are usually processed in the cloud. In Europe, processing of personal data is mainly regulated by the Directive 95/46/EC, which is currently under revision [3]. The Directive imposes quite stringent duties and obligations on the actors of such processing, mainly on the Controller but also on the Processor. Given the above, the fact that personal data can be rapidly transferred by the CSP from one datacenter to another and customer has usually no control or knowledge over the exact location of the provided resources, understandably stimulate customers concerns on data protection and data security compliance.

Many businesses that would benefit significantly from using cloud storage are holding back because of data leakage fear [4]. The cloud is a multi-tenant environment, where resources are shared. It is also an outside party, with the potential to access a customer's data. Sharing storage hardware and placing data in the hands of a vendor seem, intuitively, to be risky. Whether accidental, or due to a malicious hacker attack, data leakage would be a major security violation. The best strategy is to assume from the start that the cloud vendor is compromised and send only encrypted files to the cloud. Use the strongest encryption that one can, anything less is not worthwhile.

Data Protection in cloud can be done by access control lists to define the permissions attached to the data objects, Storage encryption to protect against unauthorized access at the data center (especially by malicious IT staff), Transport level encryption to protect data when it is transmitted, Firewalls to include web application firewalls to protect against outside attacks launched against the data

15 Improved Technique for Data Confidentiality in Cloud Environment

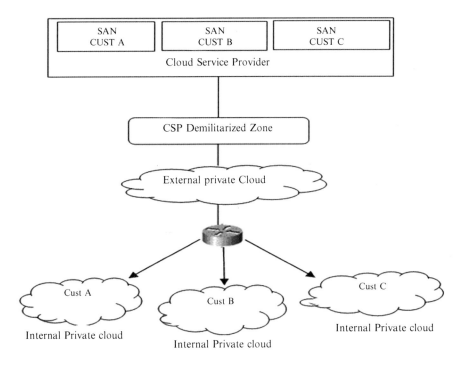

Fig. 15.1 Security component in cloud architecture [5]

center, Hardening of the servers to protect against known, and unknown, vulnerabilities in the operating system and software, Physical security to protect against unauthorized physical access to data [4].

Security is an important aspect of cloud computing as user's data security resides in remote servers which are under control of Cloud Service Providers (CSPs). Google Drive and Amazon S3 cloud services etc. are also prone to security threats. Security issues associated with cloud computing are divided into two categories: Security issues faced by cloud service providers and security issues faced by customers. Providers must ensure that their infrastructure is secure and their client, data and applications are protected while the customer must ensure provider has taken sufficient measures to protect data. Customer uses the cloud services such as Amazon EC2, Amazon S3, Google Drive, Skydrive, SugarSync, Dropbox to store bulk of their data. Customer data security is a primary concern of cloud service providers. Customer's data should be protected from the unauthorized users. Data leakage may be caused by external attackers and in some, CSP themselves may breach the user.

Security components in cloud architecture [5] are shown in Fig. 15.1. Security profile must be defined at each level. CSP must keep the infrastructure and software behind a demilitarized zone (DMZ). Operating Systems and Virtualization are handled behind the DMZ on a CSP. CSP handle resource provisioning by separating and isolating VM resources. Network Security is provided through router ACLs, perimeter firewall or web application security. CSP provide access paths to physical servers that have permissions for the desired functionality. Authentication, authorization and auditing (AAA) must be provided by the CSP.

In this paper, we propose an improved ELGamal Encryption/Decryption algorithm. This algorithm will be implemented in the CSP servers. Whenever user signs up, a pair of public key and private key is created for the user. When user login and want to store the file in the cloud, CSP will encrypt the file in the users' web browser and stores the file in the cloud. When user wants to access the encrypted file, CSP will return the file to user and decrypt the file using user's private key in the users' web browser. This algorithm is better both in terms of space and time than earlier ELGamal algorithm [6]. In the existing ELGamal algorithm, each character was replaced with two characters, but we modified the algorithm and now two characters are replaced with three characters. It achieves better space and time complexity than the existing algorithm.

The remaining paper is organized as follows: Related work about data security in cloud computing is surveyed in Sect. 15.2. Section 15.3 describes our proposed technique which compares with the existing ELGamal algorithm. Section 15.4 describes experiments conducted by encrypting many files and observing the results in terms of time and space. We conclude our paper in Sect. 15.5 while giving directions for future work.

15.2 Related Work

Yobu Tan presented a cloud computing data security solutions both for the save transmission and storage of data [6]. He talked about full Homomorphic encryption technique and its features. He proposed a technique that uses user's public key to encrypt the data. When the data is encrypted, it is stored in the cloud. When user wants to access the file, it returns file to user. User decrypts it by using its own private key. It reads one character at a time from the plain text file and encrypts it into two characters. The execution time is high to process the file. The existing algorithm just doubles the size of plain text when it is encrypted.

Christodorescu et al. proposed a scalable solution that centralizes guest protection into a security VM. It supports Linux and Windows operating systems and can be easily extended to support new operating systems. It does not assume any a-priori semantic knowledge of the guest and does not require any a-priori trust assumptions into any state of the guest VM [7]. They focused their current implementation of anti-rootkit engine to kernel-level malware since excessive overhead is involved in monitoring in the user interface of the guest VM. Expensive context switching is required to monitor too many events. They plan to address the limitation through the injection into the guest VM of security agents, which would run locally to identify user interface malware.

Xiao et al. states recent advances have given rise to the popularity and success of cloud computing [8]. However, when outsourcing the data and business application to a third party causes the security and privacy issues to become a critical concern. Throughout the study at hand, the authors obtain a common goal to provide a comprehensive review of the existing security and privacy issues in cloud environments. They have identified five most representative security and privacy attributes (i.e. Confidentiality, Integrity, Availability, accountability, and privacy-preservability). Beginning with these attributes, they present the relationships among them, the vulnerabilities that may be exploited by attackers, the threat models, as well as existing defense strategies in a cloud scenario. Future research directions are previously determined for each attribute.

Zhao aims to construct a system for trusted data sharing through untrusted cloud providers to address the above mentioned issue [9]. The constructed system can imperatively impose the access control policies of data owners, preventing the cloud storage providers from unauthorized access and making illegal authorization to access the data. He proposes a progressive encryption scheme based on elliptic curve encryption. The proposed progressive encryption scheme allows data to be encrypted multiple times with different keys and produces a final cipher text that can be decrypted with a single decryption key in a single decryption operation. This scheme allows changing the encryption key without decrypting the data first, thus enables the re-encryption of data in an untrusted environment. He devised a scheme for secure sharing on the cloud. The protocol is devised based on the proposed progressive encryption scheme, allowing a data owner to store its encrypted data on a cloud and share with different users. The sharing is achieved by re-encrypting the data to the authorized users by the cloud provider.

Prasad et al. proposes a framework that works in two stages [10]. First stage is the Data classification which is done by client before storing the data. Data is categorized on the basis of CIA (Confidentiality, Integrity, and Availability) during this stage. The client who wants to upload the data in cloud gives the value of C (Confidentiality), I (Integrity), A (Availability). The value of C is based on level of secrecy at each point of data processing and prevents unauthorized disclosure, value of I based on how much assurance of accuracy is provided, reliability of information and unauthorized modification is required, and value of A is based on how frequently it is accessible. The priority rating is calculated by using proposed formula. The data critical is one with higher rating and 3D security is recommended on that data. Cloud provider uploads the data after the first phase and uses 3 Dimensional techniques for accessing the data. The sensitive proved data will send for storage to cloud provider. According to the concept of 3D user who wants to access the data need to be authenticated, to avoid impersonation and data leakage. Now there is third entity who is either company's (whose data is stored) employee or customer who want to access, they need to register first and then before every access to data, his/her identity is authenticated for authorization.

Sood et al. proposed a framework comprising of different techniques and specialized procedures is proposed that can efficiently protect the data from the beginning to the end, i.e. from the owner to the cloud and then to the user [11].

We commence with the classification of data on the basis of the cryptographic parameters presented by the user i.e., Confidentiality (C), Availability (A) and Integrity (I). The strategy followed to protect the data utilizes various measures such as the SSL (Secure Socket Layer) 128-bit encryption and can also be raised to 256-bit encryption if needed, MAC (Message Authentication Code) is used for integrity check of data, searchable encryption and division of data into three sections in cloud for storage. The division of data into three sections renders supplementary protection and simple access to the data. The user who wishes to access the data is required to provide the owner log in identity and password, before admittance is given to the encrypted data.

Zissis et al. states security requirements and provides the viable solution that eliminates potential threats [12] to cloud security. TTP is an intermediate trusted third party between two entities of different administrative domains to establish secure intersection. A TTP addresses a number of security issues in a multilevel distributed environment. The proposed solution calls upon cryptography, specifically Public Key Infrastructure operating in concert with SSO and LDAP, to ensure the authentication, integrity and confidentiality of involved data and communications. The solution, presents a horizontal level of service, available to all implicated entities, that realizes a security mesh, within which essential trust is maintained.

15.3 Proposed Technique

ELGamal encryption perform the following steps: first of all, the key generator generates two keys (public key and private key), then the encryption algorithm encrypts the data by public key of user, and the decryption algorithm decrypts the data by using private key of user when it is needed. We studied existing ELGamal technique for encryption of data in cloud. We modified this technique to provide better time and space complexity in encryption. Section 15.3.1 describes the older technique and Sect. 15.3.2 describes the proposed technique.

15.3.1 ELGamal Encryption Algorithm

ELGamal algorithm (refer to Fig. 15.2) is a public key cryptography algorithm that uses an asymmetric technique. It reads the data character by character and converts each character into two characters i.e. each character is encrypted into two characters. First of all a key generator function generates a set of public and private keys i.e. (y, q, p) and (X). Then sender sends the data to CSP. Then CSP encrypts each character into two characters and stores them in the cloud. When sender wants to access the data from cloud, CSP reads two encrypted characters and decrypts them into one character.

```
Choose two large prime number p & q(q<p) from the
cyclic group of Z*p of the generator.
Choose a random no X ∈ Z*p
Compute y = q^x mod p
public key   : (y, q, p)
private key  : (X)
```
Encryption:
```
Choose a random no k and p-1 are relatively prime
     E(M) = (a, b)
          = (q^K mod p, y^K M mod p)
```
Decryption:
```
     D(M) = b(a^X)^-1 mod p
```

Fig. 15.2 ELGamal algorithm

```
Choose two large prime number p & q (q<p) from
the cyclic group of Z*p of the generator
Choose a random no X ∈ Z*p
Compute y = q^x mod p
Public key  : (y, q, p)
Private key : (X)
```
Encryption:
```
Choose a random no k and p-1 are relatively
prime
    E(M1, M2) = (a, b, c)
              = (q^K mod p, y^K M1 mod p, y^K M2 mod p)
```
Decryption:
```
        D(M1) = b(a^X)^-1 mod p
        D(M2) = c(a^X)^-1 mod p
```

Fig. 15.3 Proposed technique

15.3.2 Modified ELGamal Algorithm

Proposed technique (refer to Fig. 15.3) is an asymmetric key encryption algorithm for public key cryptography. Unlike ELGamal algorithm, the proposed technique reads two characters at a time and converts them into three characters. First of all a key generator function generates a set of public and private keys i.e. (y, q, p) and (X). Then sender sends the data to CSP. Then CSP encrypts each two character into three characters and stores them in the cloud. When sender wants to access the data from cloud, CSP reads three encrypted characters and decrypts them into two characters.

Implementation steps are as follows (refer to Fig. 15.4):

- User Sign up to cloud, CSP produces the public key and private key.
- Whenever the user wants to upload the data on the cloud, then first he logins and then the system selects the user's public key to encrypt data, and the encrypted

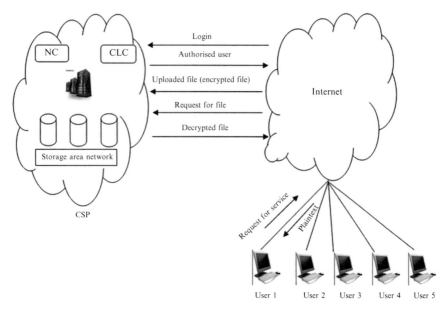

Fig. 15.4 File storage and retrieval process

data is uploaded to the cloud. The key feature of this algorithm is that this algorithm is reading the file character by character. It reads two characters at a time and encrypt into three characters. Before reading next two characters form the file, the prior encrypted characters are saved on the cloud. In this way, security is incorporated to the cloud.

- When the users wants to access data from the cloud, he logins to the cloud. At this point, the user sends an encrypted request to the cloud and then CSP uses the private key of the user to decrypt the encrypted data, and return the results to the user.

15.4 Experiments and Results

The algorithms are analyzed based on the amount of resources such as time and storage needed to execute them. The efficiency of an algorithm is measured in terms of the time complexity and space complexity. The proposed technique takes the less time and less space over same data set. For analysis of two techniques, we run the two techniques over some dataset of different size by using Java.

15 Improved Technique for Data Confidentiality in Cloud Environment

Table 15.1 Space complexity

File name	File size in MB	Existing algorithm encrypted file size in MB	Proposed algorithm encrypted file size in MB	Improvement (%)
File 1	1.04	2.08	1.56	25
File 2	1.11	2.22	1.66	25
File 3	1.44	2.88	2.16	25
File 4	2.42	4.84	3.64	25
File 5	2.67	5.35	4.01	25
File 6	6.45	12.9	9.68	25

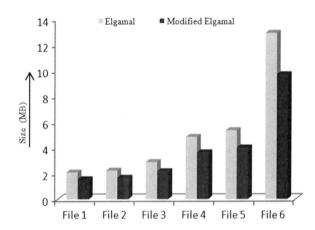

Fig. 15.5 Space complexity

15.4.1 Space Complexity

A measure of the amount of memory required to execute an algorithm with respect to the input size. Space complexity includes both Auxiliary space and space used by input. We measured the encrypted file size by running both the techniques over some plain text files of different size. It can be observed from Table 15.1 and Fig. 15.5 that Technique A is doubling the size of plain text file when it is encrypted because it replaces the one character by two characters. So the encrypted file size is 2* plain text file size. Hence less efficient Whereas Technique B is an improvement over the Technique A. Here we select two characters and encrypt them into three characters, so the encrypted file size is 3/2*plain text file i.e. 1.5 % greater than the original file. If users wishes to storage larger amount of data, there will be significant saving in the storage in case of Technique B.

Table 15.2 Time complexity

File name	File size in MB	Existing algorithm execution time in s	Proposed algorithm execution time in s	Improvement (%)
File 1	1.04	35	18	48
File 2	1.11	39	30	23
File 3	1.44	53	40	33
File 4	2.42	60	48	20
File 5	2.67	98	82	16
File 6	6.45	253	196	22

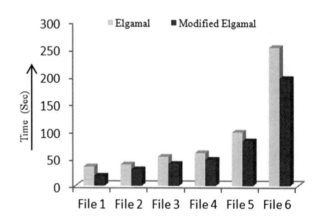

Fig. 15.6 Time complexity

15.4.2 Time Complexity

A measure of the amount of time required to execute an algorithm. We run our algorithm in different files of different size and we get different execution time. We run the existing technique also on all those files and noted the execution time. It can be seen from Table 15.2 and Fig. 15.6 that the execution time is much better than the existing technique in all different dataset. The program's run-time is directly proportional to its input size. Doubling the input size doubles the run time, quadrupling the input size quadruples the run-time, and so on. On the other hand, Program B, Doubling the input size only increases the run time by a constant amount.

15.5 Conclusion

This paper provides an encryption scheme for the secure data storage on cloud that will prevent unauthorized users to access others data stored on the cloud. The proposed technique is an improvement over the ELGamal algorithm.

The algorithmic analysis described in Sect. 15.4 proves that proposed technique has better time and space complexity than ELGamal encryption technique.

In future, we will incorporate a compression technique in our proposed encryption technique, which will make our technique more secure and will save the storage needed by CSP to store the encrypted data files. There will be significant reduction in space needed for storage of user's data.

References

1. R. Buyya, C. Vecchiola, S.T. Selvi, *Mastering Cloud Computing* (McGraw-Hill, New Delhi, 2013)
2. The state of cloud security report. http://www.alertlogic.com/resources/cloud-security-report/
3. A.D.P. Balboni, Data protection and data security issues related to cloud computing in the EU. in *ISSE 2010 Securing Electronic Business Processes*, Springer, 2010, pp. 163–172. http://dl.frz.ir/cs/239%20Computer%20Science%20Springer%20Books%20[Guerrilla%20Open%20Access]/Norbert%20Pohlmann,%20Helmut%20Reimer,%20Wolfgang%20Schneider%20-%20ISSE%202010%20Securing%20Electronic%20Business%20Processes.pdf
4. Top 5 security challenges of cloud storage. http://www.nasuni.com/news/press_releases/26-top_5_security_challenges_of_cloud_storage
5. A.K. Mishra, P. Matta, E.S. Pilli, R.C. Joshi, Cloud forensics: state-of-the-art and research challenges. in *International Symposium on Cloud and Services Computing (ISCOS)*, 2012, pp. 164–170. http://ieeexplore.ieee.org/xpl/articleDetails.jsp?tp=&arnumber=6481255&queryText%3Dstate-of-the-art+and+research+challenges
6. T. Yubo, W. Xinlei, Research of cloud computing data security technology. in *2nd International Conference on Consumer Electronics, Communications and Networks (CECNet)*, 2012, pp. 2781–2783. http://www.ijert.org/browse/volume-2-2013/february-2013-edition?download=2282%3Adata-a-service-and-their-security-concerns-in-cloud&start=10
7. M. Christodorescu, R. Sailer, D.L. Schales, D. Sgandurra, D. Zamboni, Cloud security is not (just) virtualization security. in *Proceedings of the 2009 ACM workshop on Cloud Computing Security*, ACM, 2009, pp. 97–102. http://dl.acm.org/citation.cfm?id=1655022
8. Z. Minqi, Z. Rong, X. Wei, Q. Weining, Z. Aoying, Security and privacy in cloud computing: a survey. in *Sixth International Conference on Semantics Knowledge and Grid (SKG)*, 2010, pp. 105–112. http://www.ccsenet.org/journal/index.php/nct/article/download/21403/14405
9. Z. Gansen, R. Chunming, L. Jin, Z. Feng, T. Yong, Trusted data sharing over untrusted cloud storage providers. in *IEEE Second International Conference on Cloud Computing Technology and Science (CloudCom)*, 2010, pp. 97–103. http://ieeexplore.ieee.org/xpl/articleDetails.jsp?tp=&arnumber=5708439&queryText%3DTrusted+data+sharing+over+untrusted+cloud
10. P. Prasad, B. Ojha, R.R. Shahi, R. Lal, A. Vaish, U. Goel, 3 dimensional security in cloud computing. in *3rd International Conference on Computer Research and Development (ICCRD)*, 2011, pp. 198–201. http://ieeexplore.ieee.org/xpl/articleDetails.jsp?tp=&arnumber=5764279&queryText%3D3+dimensional+security+in+cloud+computing
11. S.K. Sood, A combined approach to ensure data security in cloud computing. J. Netw. Comput. Appl. **35**, 1831–1838 (2012)
12. D. Zissis, D. Lekkas, Addressing cloud computing security issues. Elsevier **28**, 583–592 (2012)

Chapter 16
A Hybrid-Based Feature Selection Approach for IDS

Amrita and P. Ahmed

Abstract An intrusion detection (ID) technique classifies the incoming network traffic, represented as a feature vector, into anomalous or normal traffic by a classification method. In practice, it has been observed that the high dimensionality of the feature vector degrades classification performance. To reduce the dimensionality, without compromising the performance, a new hybrid feature selection method has been introduced and its performance is measured on KDD Cup'99 dataset by the classifiers Naïve Bayes and C4.5. Three sets of experiments have been conducted using full feature set, reduced sets of features obtained using four well known feature selection methods as Correlation-based Feature Selection (CFS), Consistency-based Feature Selection (CON), Information Gain (IG), Gain Ratio (GR) and the proposed method on the said dataset and classifiers. In first experiment, classifier Naïve Bayes and C4.5 yielded classification accuracy 97.5 % and 99.8 % respectively. In second set of experiments, the best performance (accuracy) of these classifiers was achieved as 99.1 % and 99.8 % by the method IG. In third experiment, six features are obtained using proposed method and noted the same as 99.4 % and 99.9 %. The proposed hybrid feature selection method outperformed earlier mentioned methods on various metrics.

16.1 Introduction

In net-centric computing environments a security infringement is considered as an intrusion. The spreading net-centric computing, like cloud computing, and increasing intrusion problems have created a dire need for an efficient and reliable Intrusion Detection System (IDS). An IDS monitors computer network traffic, identifies malicious traffic, i.e., the traffic that may harm data and software, and

---

Amrita (✉) · P. Ahmed
Department of CSE, SET, Sharda University, Greater Noida, India
e-mail: amrita.prasad@sharda.ac.in; pervez.ahmed@sharda.ac.in

raises alerts about such traffics. The IDS employs statistical pattern techniques to classify an unknown network traffic into a regular (normal) or an anomalous (abnormal) class. This anomalous traffic is either denied to access or processed further for discovering the anomaly types. In pattern recognition, the former is a two-class, i.e., the binary class problem. The latter is multi-class problem and it needs to discover intrusion types.

The scope of this work is limited to the former type only – which is a binary-class problem. In practice, feature vector dimensionality curse affects the performance of every pattern recognition problem. Therefore, in here, we have focused on discovering an optimal feature subset, by removing the irrelevant and redundant features, for intrusion classification. In this work, we considered high recognition accuracy and least possible cardinality of the feature subset as optimality criteria [1]. The optimal feature set guarantees high true positive (TP) rate, low false positive (FP) rate, less time to build model and minimum errors.

In an attempt to select an optimal set, a novel hybrid feature selection method has been developed and its performance has been tested using binary classifier on non-redundant discretized data of 10 % of KDD Cup'99 dataset. The performance in terms of TP rate, FP rate, time to build model and various types of error as mean absolute error (MAE), root mean squared error (RMSE), relative absolute error (RAE) and root relative squared error (RRSE) are used.

The rest of the paper is organized into the following sections. Related work is reviewed in Sect. 16.2. Section 16.3 gives the description about the feature selection method and classifiers used in this work. Dataset used in this paper is discussed in Sect. 16.4 and the proposed feature selection algorithm in Sect. 16.5. Experimental setup and results in Sects. 16.6 and 16.7. Section 16.8 concludes and discusses future work.

16.2 Related Work

Filter and wrapper methods were used in a work by Wang et al. [2]. In their work, Information gain for filter method and Bayesian Networks (BN) and decision trees (C4.5) for wrapper method were employed to select features for network intrusion detection. Ten features were selected by using this method. Detection rate and false positive rate by classifier C4.5 are 99.80 % and 0.26 % respectively. Another, a new hybrid approach named as C4.5-PCA-C4.5 was proposed in [3]. It uses PCA (Principal Component Analysis) and decision tree classifier C4.5 as feature selection method and C4.5 as classifiers. Seven important features were selected and average building process time for C4.5-PCA-C4.5 was 6 seconds. A new approach to model lightweight Intrusion Detection System (IDS) based on a new feature selection approach named Correlation-based Hybrid Feature Selection (CBHFS) which was able to significantly decrease training and testing times while retaining high detection rates with low false positives rates as well as stable feature selection results proposed by Park et al. [4].

Fusion of Genetic Algorithm (GA) and Support Vector Machines (SVM) was also proposed in [5] for efficient optimization of both features and parameters for detection models. This method was proved to be an efficient way of selecting important features as well as optimizing the parameters for detection model and provides more stable detection rates. A fast hybrid feature selection method to determine an optimal feature set was proposed in [6]. This method was a fusion of Correlation-based Feature Selection (CFS), Support Vector Machine (SVM) and Genetic Algorithm (GA). Twelve features were selected and achieved 99.56 % as TP rate and 37.5 % as FP rate in average.

16.3 Feature Selection Methods and Classifiers

16.3.1 Feature Selection Methods

Feature selection is the selection of that minimal dimensionality feature subset of original feature set that retains the high detection accuracy as the original feature set [1]. Blum and Langley [7] divided the feature selection methods into three categories named filter [8], wrapper [9] and hybrid [10] (embedded) method. Filter method uses external learning algorithm to evaluate the performance of selected features. The wrapper method "wrap around" the predefined classifier to evaluate subsets of features. The hybrid method combines the wrapper and the filter method to achieve the best possible performance with a particular learning algorithm. The paper [11] presented a survey of various feature selection methods for IDS on KDD CUP'99 dataset based on these three categories and different evaluation criteria. The proposed method is based on hybrid method. We selected four well-known filter based feature selection algorithms as Correlation-based Feature Selection (CFS), Consistency-based Feature Selection (CON), Information Gain (IG) and Gain Ratio (GR) for this work.

The CFS [12] is a filter method. It assumes that an optimal feature subset contains feature elements that are highly correlated with the classification and are uncorrelated with each other. The CON [13] is also a filter method. It evaluates the worth of a subset of features by the level of consistency in the class values when the training instances are projected onto the subset of attributes. This method uses a consistency measures to find the smallest set of features with consistency equal to that of the full set of features. The IG [14] based feature selection is a feature ranking method. It evaluates features by measuring their information gain with respect to the class. The GR [15] is also a method of feature ranking for feature selection. It is an extension of information gain and attempts to overcome the bias.

Table 16.1 Lists of feature number (#) and corresponding name in the KDD Cup'99

#	Name	#	Name	#	Name
1	Duration	15	Su-attempted	29	Same-srv-rate
2	Protocol-type	16	Num-root	30	Diff-srv-rate
3	Service	17	Num-file-creations	31	Srv-diff-host-rate
4	Flag	18	Num-shells	32	Dst-host-count
5	Src-bytes	19	Num-access-files	33	Dst-host-srv-count
6	Dst-bytes	20	Num-outbound-cmds	34	Dst-host-same-srv-rate
7	Land	21	Is-hot-login	35	Dst-host-diff-srv-rate
8	Wrong-fragment	22	Is-guest-login	36	Dst-host-same-src-port-rate
9	Urgent	23	Count	37	Dst-host-srv-diff-host-rate
10	Hot	24	Srv-count	38	Dst-host-serror-rate
11	Num-failed-logins	25	Serror-rate	39	Dst-host-srv-serror-rate
12	Logged-in	26	Srv-serror-rate	40	Dst-host-rerror-rate
13	Num-compromised	27	Rerror-rate	41	Dst-host-srv-rerror-rate
14	Root-shell	28	Srv-rerror-rate		

16.3.2 Classifiers

Two supervised machine learning classifiers: Naïve Bayes [16] and C4.5 [17] are used in this work. These two classifiers are selected because they can work on both numerical and symbolic features. They are also faster and more computationally efficient. Naïve Bayes is simple supervised learning classifier based on Bayes' theorem of probability theory. It is widely used classifier having many properties and is highly suitable for high dimensional large dataset [16]. The C4.5 algorithm is also very robust for high dimensional large data and handling missing data.

16.4 KDD Cup 1999 Dataset

The KDD CUP 1999 [18] is a benchmark dataset for IDS. It contains 4,940,000 and 311,029 connection records for training data set and test data set respectively. Since the training set is prohibitively large, 10 % of the KDD Cup'99 dataset is chosen as experimental dataset. It contains 494,021 connection records in which 97,277 are normal and 396,744 are attack. Each connection has a label of either normal or the attack type. The attack type falls into one of the four attack categories [19] as: Denial of Service Attack (DoS), User to Root Attack (U2R), Remote to Local Attack (R2L) and Probing Attack. Each connection record consisted of 41 features plus one class label. These 41 features are labeled in order as 1, 2, 3, 4, 5, 6, 7, 8, 9,..., 41 shown in Table 16.1. There are 32 numerical features (continuous values) and 9 symbolic features (discrete).

16.5 The Proposed Method for Feature Selection

We proposed a hybrid method for feature selection by using wrapper method with the classifier Naïve Bayes and fusion of filter based feature selection methods. Four filter based feature selection methods CFS, CON, IG and GR are selected for fusion. From these four feature selection methods, the CFS and CON select the subset of features from the given set of features while IG and GR rank the individual feature according to its relevance.

First, we obtained the initial feature subset by investigating the fusion of four filter based feature selection methods. This initial feature subset is created by selecting the relevant features present among the feature sets obtained by applying the methods CFS with best first search, CON with best first search, IG with ranker and GR with ranker. Since IG and GR rank the features therefore we arranged the features in descending order of its rank. Then, we choose the first N1 features from IG and first N2 features from GR based on performance. In this process, first we take the common features from the feature sets of CFS and CON. Similarly, the common features from N1 of IG and N2 of GR are selected. These two common feature subsets obtained are added to create the initial feature subset. Another set is a set of left features which are left in the sets of CFS, CON, N1 of IG and N2 of GR. Further, wrapper based feature selection method with Naïve Bayes classification algorithm is employed to find the final optimal feature subset. Linear Forward selection (LFS) is adapted in the wrapper based feature selection. In the proposed approach, LFS starts with the initial feature subset and then add features one by one from the left feature set until there is no change in the performance of current subset by adding features. The performance in terms of TP rates, FP rate, time to build the model and errors are used to select the final optimal feature set. The detail procedure of proposed algorithm is described by the following steps:

Algorithm for proposed method.

Input: Discretized Non-redundant data of 10% KDD Cup'99 dataset (Section 6.1)

Output: An optimal set of features.

Method:

Step 1: Initialize $F_{full} = \{f_1, f_2, f_3, \ldots, f_{41}\}$ is the set of full 41 features of dataset.

Step 2: Obtain feature set by applying CFS (CfsSubsetEval+Best First) on F_{full} set. Let F_{CFS} be the set.

(continued)

Algorithm for proposed method (continued)

Step 3: Obtain feature set by applying CON (ConsistencySubsetEval+BestFirst) on F_{full} set. Let F_{CON} be the set.

Step 4: Obtain feature's rank by applying IG on F_{full} set. These features are then arranged in descending order based on their rank. Top N1 features are selected from the arranged list based on the performance. Let F_{IG} and $F_{IG(N1)}$ be the set.

Step 5: Obtain feature's rank by applying GR on F_{full} set. These features are then arranged in descending order based on their rank. Top N2 features are selected from the arranged list based on the performance. Let F_{GR} and $F_{GR(N2)}$ be the set.

Step 6: Obtain features that are common in the sets obtained at steps (2) and (3). Let $F_{CFS \cap CON}$ be the set of features. Similarly, obtain the common features from the sets obtained at steps (4) and (5). Let $F_{IG(N1) \cap GR(N2)}$ be the set of features.

$$F_{CFS \cap CON} = (F_{CFS} \cap F_{CON}); \quad F_{IG(N1) \cap GR(N2)} = \left(F_{IG(N1)} \cap F_{GR(N2)}\right)$$

Step 7: Add the two sets $F_{CFS \cap CON}$ and $F_{IG(N1) \cap GR(N2)}$ obtained at step 6. This set is the initial set of features used as the initial set for wrapper method with the classifier. Let F_{IniFea} be the set.

$$F_{IniFea} = F_{CFS \cap CON} \cup F_{IG(N1) \cap GR(N2)}$$

Step 8: Now consider the remaining features from the four sets F_{CFS}, F_{CON}, $F_{IG(N1)}$ and $F_{GR(N2)}$ that are left after selecting the feature set obtained at step 7. Let F_{Left} be the set of left features.

$$F_{\cup} = F_{CFS} \cup F_{CON} \cup F_{IG(N1)} \cup F_{GR(N2)}; \quad F_{Left} = F_{\cup} - F_{IniFea}$$

(continued)

Algorithm for proposed method (continued)

Step 9: $F_{Temp} = F_{IniFea}$ = set of initial features obtained at step 7. F_{Left} = set of left features obtained at step 8. Per_{Temp} = Performance of F_{Temp}, p = number of features in F_{Left}, $F_{Current}$ = Empty set.

```
for i=1 to p
begin
  Select i^th feature from F_Left set.
  F_Current = F_Temp ∪ F_Left_i
  Compute performance of F_Current as Per_Current by
    classifier
  if Per_Current > Per_Temp then
    F_Temp = F_Temp ∪ F_Left_i
  else
    F_Current = F_Temp
  end if
end for
```

Step 10: The set F_{Temp} is the final optimal feature set. Let the final set of features be F_{Final}. Test the performance of the sets F_{CFS}, F_{CON}, $F_{IG(N1)}$, $F_{GR(N2)}$, F_{IniFea} and F_{Final} by the classifier.

The experimental setup and results for proposed method are described Sects. 16.6 and 16.7.

16.6 Experimental Setup

WEKA [20], a machine learning tool is used to compute the different feature sets by CFS with best first search (CfsSubsetEval + BestFirst), CON with best first search (ConsistencySubsetEval + BestFirst), IG with rank (InfoGainAttributeEval + Ranker), and GR with rank (GainRatioAttributeEval + Ranker) and for classifiers Naïve Bayes and C4.5 to measure the classification performance for different feature sets obtained at different steps of proposed method. The steps followed to implement the proposed method are preprocessing of dataset, feature selection and feature evaluation.

16.6.1 Preprocessing of Dataset

"10 % KDD Cup'99" dataset are preprocessed for experiment. For binary class (attack or normal) classification problem, the label of each connection must have a label of either normal or the attack. Therefore, all the types of attack are converted into label "attack". This dataset also contains redundant connection records. These redundant records will influence the performance of the classifiers. After removing these redundant records, the resultant dataset is reduced to 145,586 from 494,021. Around 70 % redundant records are there in this dataset. Further, feature selection methods and classifiers used in this paper work on discrete data. We used Entropy Minimization Discretization method proposed by Fayyad and Irani [21] for discretization. The resultant discretized dataset is *"Dis_Non-Redundant 10 % KDD Cup'99"*.

16.6.2 Feature Selection

Feature subset selection is done according to the proposed method stated in Sect. 16.5. We used Naïve Bayes classifier to test the different feature sets obtained at different steps of proposed method and also in wrapper method. The rank of features obtained by IG and GR are shown in Tables 16.2 and 16.3 respectively. The features in Tables 16.2 and 16.3 are arranged in descending order based on their ranks. From Tables 16.2 and 16.3, it can be observed that features {7, 9, 14, 15, 18, 20, 21, 22} are common to both tables having rank 0. Therefore there are only 33 features which are relevant for further processing. The TP rate, FP rate, and RMSE of first N number of features, where N is varying from 1 to 33, from Tables 16.2 and 16.3, are obtained by classifier Naïve Bayes. The number of features selected based on obtained result for IG and GR are 3 and 25 respectively. Therefore values of N1 and N2 are 3 and 25 at step 4 and 5 of proposed method respectively. Selected feature sets and performance of the feature sets obtained at each step by the proposed method are shown in Table 16.4. Finally six important features are selected by the proposed method.

16.6.3 Feature Evaluation

Naïve Bayes and C4.5 classifiers are used for feature evaluation. The optimal feature set obtained in feature selection phase is evaluated by Naïve Bayes and C4.5 classifiers. The performance of optimal feature set is tested in terms of TP rate, FP rate, time to build model and various types of error. For experiments, we used tenfold cross validation to evaluate the different feature sets by classifiers.

Table 16.2 Rank of feature obtained by InfoGainAttributeEval + Ranker

S. no	Feature#	Rank	S. no	Feature#	Rank	S. no	Feature#	Rank	S. no	Feature#	Rank
1	5	0.777642	12	39	0.53976	23	40	0.036751	34	18	0
2	3	0.770687	13	25	0.53894	24	27	0.0359	35	14	0
3	29	0.76049	14	26	0.535155	25	28	0.034621	36	15	0
4	30	0.758437	15	12	0.471499	26	2	0.033233	37	22	0
5	23	0.724922	16	36	0.311277	27	10	0.010456	38	9	0
6	4	0.694624	17	37	0.304817	28	8	0.010336	39	20	0
7	6	0.65357	18	32	0.301177	29	13	0.007595	40	7	0
8	34	0.64756	19	31	0.163721	30	16	0.002307	41	21	0
9	33	0.646073	20	24	0.108046	31	19	0.001838			
10	35	0.628309	21	41	0.069915	32	17	0.000493			
11	38	0.544408	22	1	0.038766	33	11	0.000368			

Table 16.3 Rank of feature obtained by GainRatioAttributeEval + Ranker

S. no	Feature#	Rank	S. no	Feature#	Rank	S. no	Feature#	Rank	S. no	Feature#	Rank
1	30	0.6027	12	33	0.252	23	41	0.0811	34	22	0
2	26	0.5641	13	23	0.251	24	1	0.0719	35	9	0
3	25	0.5599	14	34	0.2476	25	27	0.0657	36	7	0
4	29	0.5146	15	35	0.2372	26	11	0.0647	37	20	0
5	4	0.5101	16	37	0.2336	27	2	0.062	38	15	0
6	39	0.4781	17	36	0.1703	28	16	0.0611	39	14	0
7	12	0.4716	18	32	0.1659	29	19	0.0605	40	18	0
8	38	0.4536	19	8	0.1587	30	28	0.0604	41	21	0
9	3	0.317	20	31	0.1466	31	40	0.0447			
10	6	0.2763	21	13	0.1239	32	24	0.0308			
11	5	0.2701	22	10	0.096	33	17	0.0257			

Table 16.4 Performance and feature set obtained at each step by proposed algorithm

Step	Feature set	Methods	No of feature	Selected features	TP rate	FP rate	RMSE	Build (s)
1	All	–	41	–	0.975	0.038	0.1556	1.44
2	F_{CFS}	CFS + BestFirst	8	3, 4, 8, 12, 13, 26, 29, 30	0.974	0.038	0.1572	0.12
3	F_{CON}	CON + BestFirst	11	1, 3, 5, 6, 13, 24, 29, 33, 34, 35, 38	0.983	0.026	0.1275	0.16
4	$F_{IG(33)}$	IG + Rank	33	5, 3, 29, 30, 23, 4, 6, 34, 33, 35, 38, 39, 25, 26, 12, 36, 37, 32, 31, 24, 41, 1, 40, 27, 28, 2, 10, 8, 13, 16, 19, 17, 11	0.975	0.038	0.1556	0.34
4	$F_{IG(3)}$	IG + Rank	3	5, 3, 29	0.991	0.012	0.0940	0.08
5	$F_{GR(33)}$	GR + Rank	33	30, 26, 25, 29, 4, 39, 12, 38, 3, 6, 5, 33, 23, 34, 35, 37, 36, 32, 8, 31, 13, 10, 41, 1, 27, 11, 2, 16, 19, 28, 40, 24, 17	0.975	0.038	0.1556	0.36
5	$F_{GR(25)}$	GR + Rank	25	30, 26, 25, 29, 4, 39, 12, 38, 3, 6, 5, 33, 23, 34, 35, 37, 36, 32, 8, 31, 13, 10, 41, 1, 27	0.976	0.037	0.1550	0.25
6	$F_{CFS \cap CON}$	–	3	3, 13, 29	0.964	0.046	0.1621	0.08
6	$F_{IG(N1) \cap GR(N2)}$	–	3	3, 5, 29	0.991	0.012	0.0940	0.08
7	F_{IniFea}	–	4	3, 5, 13, 29	0.992	0.012	0.0905	0.09
8	F_U	–	26	1, 3, 4, 5, 6, 8, 10, 12, 13, 23, 24, 25, 26, 27, 29, 30, 31, 32, 33, 34, 35, 36, 37, 38, 39, 41	0.975	0.037	0.1547	0.27
8	F_{left}	–	22	1, 4, 6, 8, 10, 12, 23, 24, 25, 26, 27, 30, 31, 32, 33, 34, 35, 36, 37, 38, 39, 41	0.971	0.044	0.1678	0.22
9	F_{Temp}	Wrapper	6	3, 5, 6, 10, 13, 29	0.994	0.008	0.0698	0.10
10	**F_{Final}**	**Proposed algorithm**	**6**	**3, 5, 6, 10, 13, 29**	**0.994**	**0.008**	**0.0698**	**0.10**

16.7 Experimental Results and Analysis

CFS, CON, IG, GR and proposed feature selection methods are performed on 41 features of *"Dis_Non-Redundant 10 % KDD Cup'99"* dataset. Features obtained by methods CFS, CON, IG, GR and proposed method are 8, 11, 3, 25 and 6 respectively. Ranks of the features obtained by IG and GR are arranged in descending order according to its rank shown in Tables 16.2 and 16.3 respectively. The stepwise procedure of proposed feature selection method is shown in Table 16.4. Table 16.4 shows the feature set obtained at each step of proposed method and also the performance in terms of TP rate, FP rate, RMSE and time to build model. The final feature set is obtained based on fusion of filter based methods and wrapper method with classifier Naïve Bayes on the basis of increased performance. The final feature set is reduced to 15 % of the original feature set. These six feature sets are evaluated by classifiers Naïve Bayes and C4.5 using tenfold cross validation shown in Tables 16.5 and 16.6 respectively. The proposed method is compared against the benchmark feature selection methods CFS, CON, IG and GR on reduced discretized dataset. There is increase in TP rate, decrease in FP rate of selected feature set of proposed method than CFS, CON, IG, GR and full feature set. The time to build the model is reduced by approximately 88 %. Result shows that C4.5 outperforms Naïve Bayes in TP rate, FP rate but higher in time to build the model. Also it can be seen from the Tables 16.5 and 16.6 that there is reduction in the errors of selected feature set by proposed method than that of CFS, CON, IG, GR and full set. The Figs. 16.1, 16.2 and 16.3 shows comparative graph for classifiers performance in terms of TP rate, FP rate and time to build model on the reduced feature sets obtained by (i) CFS + BestFirst (ii) CON + BestFirst (iii) IG + Ranker (iv) GR + Ranker (v) Proposed method and (vi) full set respectively. Results show that selected features from proposed method outperforms CSF, CON, IG, GR and full set. The Figs. 16.4, 16.5, 16.6 and 16.7 shows comparative chart for classifiers performance in terms of errors as MAE, RMSE, RAE and RRSE of selected feature set by CFS, CON, IG, GR, proposed method and full set.

16.8 Conclusion and Future Work

This paper proposes a new hybrid method for features selection for binary class problem. The number of significant features selected by proposed method is six from the original 41 features. The result achieved by these six features outperformed the four benchmark methods CFS, CON, IG and GR. The results are improved in terms of reduction in feature set and time to build model and also there is increase in TP rate and decrease in FP rate and errors. The scores achieved for TP rate are 99.4 % and 99.9 % and for FP rates are 0.8 % and 0.2 % by classifiers Naïve Bayes and C4.5 respectively. Only these 6 features instead of 41 features are sufficient to classify the coming traffic either as normal or malicious. This reduced

Table 16.5 Evaluation metrics of Naïve Bayes for different feature sets

Evaluation metrics	Total feature (41)	CFS + BestFirst (8)	CON + BestFirst (11)	IG + Ranker (3)	GR + Ranker (25)	Proposed method (6)
TP rate	0.975	0.974	0.983	0.991	0.976	0.994
FP rate	0.038	0.038	0.026	0.012	0.037	0.008
Time to build model (s)	1.44	0.12	0.16	0.08	0.25	0.10
MCC	0.949	0.947	0.964	0.982	0.950	0.988
Kappa statistic	0.9475	0.9459	0.9634	0.9822	0.9486	0.9882
Mean absolute error	0.0249	0.026	0.0181	0.0136	0.0245	0.0111
Root mean squared error	0.1556	0.1572	0.1275	0.0940	0.1550	0.0698
Relative absolute error (%)	5.2065	5.4316	3.781	2.8511	5.1270	2.3203
Root relative squared error (%)	31.8085	32.1247	26.0578	19.2152	31.688	14.2642

Table 16.6 Evaluation metrics of C4.5 for different feature sets

Evaluation metrics	Total feature (41)	CFS + BestFirst (8)	CON + BestFirst (11)	IG + Ranker (3)	GR + Ranker (25)	Proposed method (6)
TP rate	0.998	0.989	0.998	0.998	0.998	0.999
FP rate	0.002	0.014	0.002	0.002	0.002	0.002
Time to build model (s)	8.71	1.19	1.37	1.08	3.46	0.63
MCC	0.997	0.977	0.996	0.997	0.996	0.997
Kappa statistic	0.9967	0.9766	0.9965	0.9968	0.9963	0.9971
Mean absolute error	0.0024	0.017	0.0025	0.0023	0.0027	0.0020
Root mean squared error	0.0382	0.0934	0.0397	0.0352	0.0403	0.0348
Relative absolute error (%)	0.5038	3.5532	0.522	0.4883	0.5625	0.4138
Root relative squared error (%)	7.8177	19.0819	8.1185	7.2053	8.2449	7.1140

Fig. 16.1 True positive (TP) rate

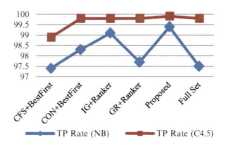

Fig. 16.2 False positive (FP) rate

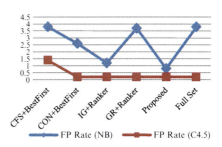

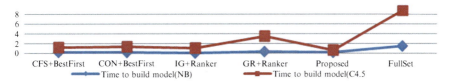

Fig. 16.3 Time to build model

Fig. 16.4 Mean absolute error

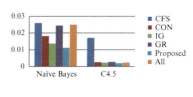

Fig. 16.5 Root mean squared error

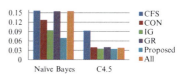

Fig. 16.6 Relative absolute error (%)

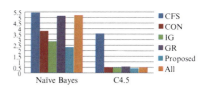

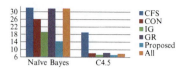

Fig. 16.7 Root relative squared error (%)

feature set also enhanced the performance and capability of IDS in real time detection and reduced the computational complexity of the classifier.

Future work will include two aspects. (i) The multiple class classification i.e. five class classification — four classes of attack (DoS, probes, U2R and R2L) and one class of normal. (ii) Improve the performance of classifiers by using the ensemble of classifiers or hybrid classifiers. This may enhance the capability and robustness of IDS.

References

1. P. Mitra et al., Unsupervised feature selection using feature similarity. IEEE Trans. Pattern Anal. Mach. Intell. **24**, 301–312 (2002)
2. W. Wang et al., Towards fast detecting intrusions: using key attributes of network traffic, in *The Third International Conference on Internet Monitoring and Protection*, 2008, pp. 86–91
3. Y. Chen et al., Building lightweight intrusion detection system based on principal component analysis and C4.5 algorithm, in *ICACT2007*, 2007, pp. 2109–2112
4. J. Park, K.M. Shazzad, D. Kim, Toward modeling lightweight intrusion detection system through correlation-based hybrid feature selection, in *Proceedings of Information Security and Cryptology*, Lecture Notes in Computer Science, Vol. 3822, 2005, pp. 279–289
5. D. Kim, H.N. Nguyen, S.Y. Ohn, J. Park, Fusions of GA and SVM for Anomaly detection in intrusion detection system, in *Proceedings of 2nd International Symposium on Neural Networks*, Lecture Notes in Computer Science, Vol. 3498, 2005, pp. 415–420
6. K.M. Shazzad, J.S. Park, Optimization of intrusion detection through fast hybrid feature selection, in *Proceedings of 6th International Conference on Parallel and Distributed Computing, Applications and Technologies*, 2005
7. A.L. Blum, P. Langley, Selection of relevant features and examples in machine learning. Artif. Intell. **97**(1–2), 245–271 (1997)
8. H. Liu, H. Motoda, *Feature Selection for Knowledge Discovery and Data Mining* (Kluwer, Boston, 1998)
9. R. Kohavi, G. John, Wrappers for feature subset selection. Artif. Intell. **97**(1–2), 273–324 (1997)
10. S. Das, Filters, wrappers and a boosting-based hybrid for feature selection, in *Proceedings of 18th International Conference on Machine Learning*, 2001, pp. 74–81
11. A. Amrita, P. Ahmed, A study of feature selection methods in intrusion detection system: a survey. Int. J. Comput. Sci. Eng. Info. Technol. Res. (IJCSEITR) **2**(3), 1–25 (2012)
12. M.A. Hall, Correlation-based feature selection for discrete and numeric class machine learning, in *Proceedings of 17th International Conference on Machine Learning*, 2000, pp. 359–366
13. M. Dash. H. Liu, Consistency-based search in feature selection. Artif. Intell. **151**(1–2), 155–176 (2003). http://dx.doi.org/10.1016/s0004-3702(03)00079-1
14. T.M. Mitchell, *Machine Learning* (Mc-Graw-Hill, New York, 1997)
15. J.R. Quinlan, Induction of decision trees. Mach. Learn. **1**(1), 81–106 (1986)
16. H. Zhang, The optimality of naive Bayes, in *The 17th International FLAIRS Conference*, Miami Beach, 2004, pp. 17–19

17. J.R. Quinlan, *C4.5: Programs for Machine Learning* (Morgan Kaufmann, San Mateo, 1993)
18. KDD Cup 1999 Intrusion detection dataset. http://kdd.ics.uci.edu/databases/kddcup99/kddcup99.html
19. S. Mukkamala et al., Intrusion detection using an ensemble of intelligent paradigms. J. Netw. Comput. Appl. **28**(2), 167–182 (2005)
20. Waikato environment for knowledge analysis (weka) version 3.7.9. Available on: http://www.cs.waikato.ac.nz/ml/weka/
21. U.M. Fayyad, K.B. Irani, Multi-interval discretization of continuous valued attributes for classification learning, in *Proceedings of the 13th International Joint Conference on Artificial Intelligence (IJCAI)*, 1993, pp. 1022–1029

Chapter 17
Reckoning Minutiae Points with RNA-FINNT Augments Trust and Privacy of Legitimate User and Ensures Network Security in the Public Network

Kuljeet Kaur and G. Geetha

Abstract Paper elucidates the process of reckoning minutiae points with reduced number of angles fingerprint algorithm. A GUI is framed to generate an evidence of improved trust and privacy of legitimate user in the public network by reckoning the bifurcations and terminations on the fingerprint of the legitimate user. After validating the authenticated user the minutiae points of the legitimate user could be exported to the database for ensuring the security of the user in the public network. Future applications for RAN-FINNT are also focused in the paper.

17.1 Introduction

Most insecure network is the Public Network in which any transaction being executed needs to have more security considerations. Intruders can attack over the legitimate users when transaction is being accomplished over the transport layer. Generally in the virtual private network of organizations SSL is implemented for enhancement in the security. But this could not completely eradicate the security threats for legitimate users. This originated the need of implementing security at the authentication level of the legitimate user. There exists lot many identity authentication parameters like Password, Smart Card and Fingerprints [1]. Usually majority of the organizations have implemented Password as most convenient identity authentication parameter. But intruders can apply password guessing attacks [2], dictionary attacks [3], stolen verifier attack [4] etc. for guessing the password of the legitimate user. There is a need to implement the most secure identity

K. Kaur (✉) · G. Geetha
School of Computer Applications, Lovely Professional University, Phagwara, India
e-mail: jeetbrar17@gmail.com

authentication parameter which is Fingerprint. It could be implemented at the entry level of the legitimate user. During its implementation extraction of the fingerprint could be possible in the form of image or texture or minutiae points [5]. As minutiae points are the set of unique points over the fingerprint and no one has the same number of minutiae points at the same place over the fingerprint [5]. This property of minutiae points makes the fingerprint of the individual unique. So in RNA-FINNT (Reduced Number of Angles Fingerprint) algorithm minutiae points are extracted at the entry level (Login Phase) for the identification of the legitimate user [5]. Implementation of RNA-FINNT for identity authentication augments trust and privacy of legitimate user and results in enhanced security over the Public Network.

Remainder sections of the paper prove this validation of RNA-FINNT for enhancing trust, privacy and security for legitimate user over the public network. Section 17.2 proposes the methodology for reckoning the minutiae points with RNA-FINNT, Sect. 17.3 demonstrates the graphical user interface as an evidence for improving trust and privacy for legitimate user, Sect. 17.4 ensures the security over the public network by exporting minutiae points to the database which will identify the legitimate user, Sect. 17.5 focuses upon the future applications of RNA-FINNT and states the references.

17.2 Process of Reckoning the Minutiae Points with RNA-FINNT

For reckoning the minutiae points the total number of bifurcations and terminations over the fingerprint are to be identified [6, 7]. The count value of bifurcations and terminations would identify the minutiae points over the fingerprint [6, 7]. Below mentioned is the process for reckoning the minutiae points:

1. Take an RGB image which would be three dimensional (Fig. 17.1a.).
2. Convert that RGB image to gray scale which would be two dimensional (Fig. 17.2).
3. Convert gray scale image to binary image which is one dimensional (Fig. 17.3).
4. Thinning of the image is done to identify exact ridges, loops and whorls over the fingerprint (Fig. 17.4).
5. Minutiae points would be found with RNA-FINNT (Fig. 17.5).
6. If false minutiae are extracted then remove those (Fig. 17.6).
7. Take a region of interest in which reckoning of bifurcations and terminations is to be done (Fig. 17.7).
8. Orientation of the fingerprint is done to verify the coordinates (Fig. 17.8).
9. Validation is done for specification of the coordinates (Fig. 17.9).
10. Reckoning of terminations is done (Fig. 17.10).

17 Reckoning Minutiae Points with RNA-FINNT Augments Trust and Privacy... 215

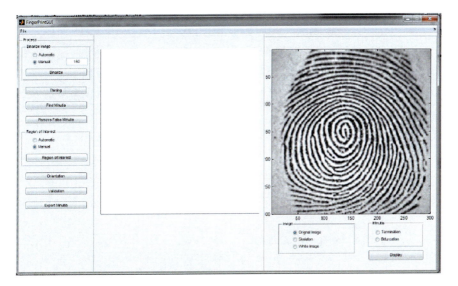

Fig. 17.1a An RGB three dimensional image

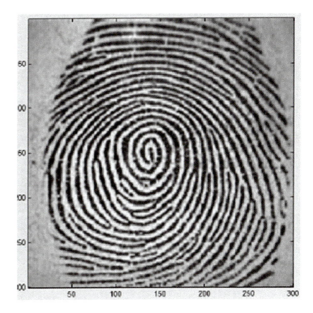

Fig. 17.1b Original RGB image

11. Reckoning of bifurcations is done (Fig. 17.11).
12. All Minutiae points could be exported to the text file and exact values of angles could be derived for ensuring the security of the legitimate user (Fig. 17.12).

13. Whenever legitimate user would login steps from 1 to 13 would be repeated and matching with the stored values would be done.
14. If match then user is legitimate otherwise malicious.

These steps would enhance the security over the public network because whenever legitimate user would login, exact reckoning value would prove the legitimacy of the user. This will fortify the trust and privacy of the legitimate user. Session could only be started if the input value would match with the exported stored value of minutiae points in the database. Program code for executing these steps:

17.2.1 Program Code for Taking an RGB Image (Fig. 17.1a)

```
function FPGUI_OpenFn(figure, userdata, vararg)
figure : for taking the image of the fingerprint
userdata : structure for handling any input through GUI
vararg : variable for inserting arguments into the GUI
```

17.2.2 Program Code for Converting RGB Image to Grayscale (Fig. 17.2)

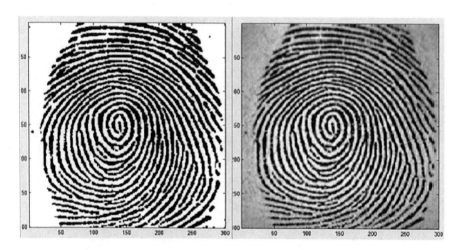

Fig. 17.2 Conversion of RGB three dimensional image to grayscale

```
setappdata(userdata.FPGUI,'BinaryImage',BinaryImage);
image(255*BinaryImage),colormap(gray)
```

17.2.3 Program Code for Converting Grayscale to Binary Image (Fig. 17.3)

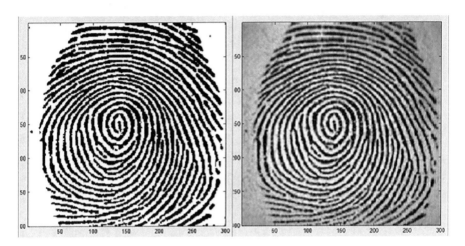

Fig. 17.3 Conversion of grayscale image to binary image

```
function Binarycall(figure, event, userdata)
I=getappdata(userdata.FPGUI,'OriginalImage');
if
      get(figure.ManualBW,'value')==1
      Threshold=str2num(get(figure.-
Threshold,'string'));
      BinaryImage=I(:,:,1)>Threshold;
else
    BinaryImage=I(:,:,1)>160;
```

17.2.4 Program Code for Thinning (Fig. 17.4)

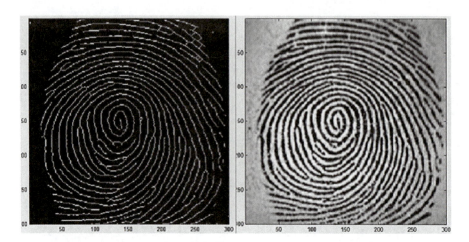

Fig. 17.4 Thinning of the original image

```
I=getappdata(userdata.FPGUI,'BinaryImage');
Skeleton=bwmorph(I,'thin','inf');
axes(figure.axes)
image(255*Skeleton)
set(gca,'tag','axes')
```

17.2.5 Program Code for Extraction of Minutiae Points with RNA-FINNT Algorithm (Fig. 17.5)

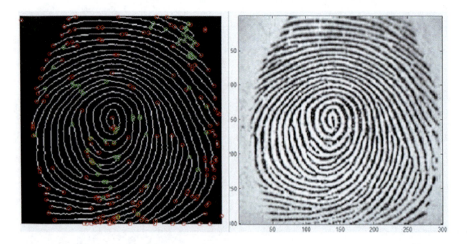

Fig. 17.5 Finding minutiae points minutiae with RNA-FINNT

```
I=getappdata(userdata.FPGUI,'Skeleton');
fun=@minutiae;
L = nlfilter(I,[3 3],fun);
Distance=DistEuclidian([CentX CentY],[CentX CentY]);
SMinutiae=Distance<D;
[i,j]=find(SMinutiae);
CentX(i)=[];CentY(i)=[];CentX(j)=[];CentY(j)=[];
```

17.2.6 Program Code for Removing False Minutiae if Extracted (Fig. 17.6)

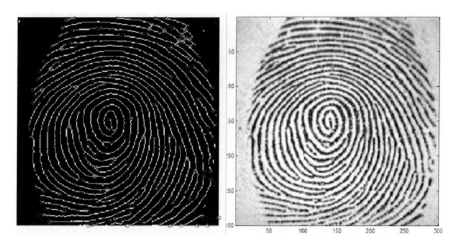

Fig. 17.6 Removal of false points

```
Distance=DistEuclidian([CentX CentY]);
SMinutiae=Distance<D;
[i,j]=find(SMinutiae);
CentX(i)=[];CentY(i)=[];
plot(CentX,CentY,'row')
plot(CentX,CentY,'Coumn')
```

17.2.7 Program Code for Taking a Region of Interest for Reckoning Terminations and Bifurcations (Fig. 17.7)

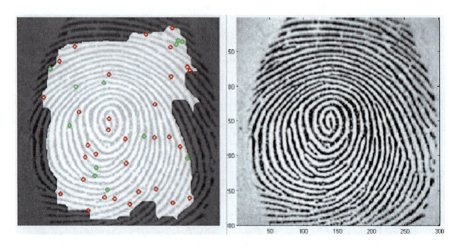

Fig. 17.7 Automatic selection of the image

```
I=getappdata(userdata.FPGUI,'OriginalImage');
S=getappdata(userdata.FPGUI,'Skeleton');
if
      get(userdata.AutomaticROI,'value')==1
      SO=imclose(S,strel('square',7));
      SOClean= imfill(SO,'holes');
      SOClean=bopen(SOClean,5);
      SOClean([1 end],:)=0;
      SOClean(:,[1 end])=0;
      ROI=imerode(SOClean,strel('disk',10));
else
      get(userdata.ManualROI,'value')==1
      f=figure;
      ROI=roipoly(I);
      close(f)
```

17.2.8 Program Code for Validating the Reckoned Terminations and Bifurcations (Figs. 17.8 and 17.9)

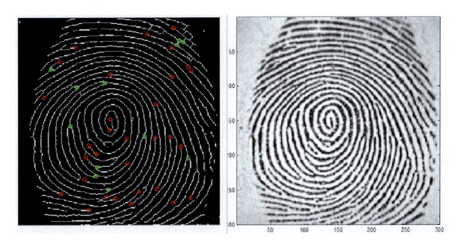

Fig. 17.8 Orientation of region of interest

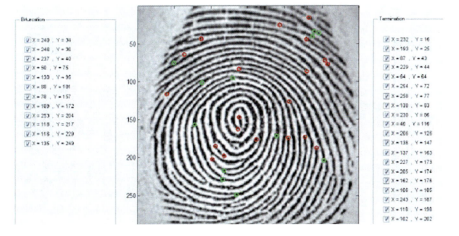

Fig. 17.9 Validation of the image

```
function Validation_Callback(figure, event, userdata)
CentX=getappdata(userdata.FPGUI,'CentX');
CentY=getappdata(userdata.FPGUI,'CentY');
CentX=getappdata(userdata.FPGUI,'CentX');
CentY=getappdata(userdata.FPGUI,'CentY');
OFin=getappdata(userdata.FPGUI,'OFin');
OSep=getappdata(userdata.FPGUI,'OSep');
I=getappdata(userdata.FPGUI,'OriginalImage');
ValidationGUI(I,CentX,CentY,OFin,CentX,CentY,OSep);
```

17.2.9 Program Code for Reckoning the Terminations (Fig. 17.10)

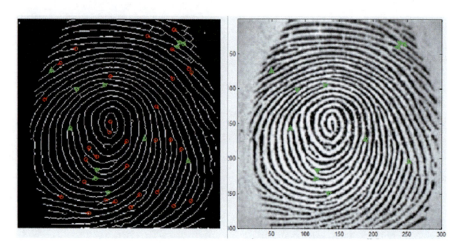

Fig. 17.10 Reckoning of terminations

```
TFin=(L==1);
TFinLab=bwlabel(TFin);
propFin=regionprops(TFinLab,'Centroid');
CentFin=round(cat(1,propFin(:).Centroid));
CentFinX=CentFin(:,1);
CentdFinY=CentFin(:,2);
axes(handles.axes)
plot(CentFinX,CentFinY,'row')
```

17.2.10 Program Code for Reckoning the Bifurcations (Fig. 17.11)

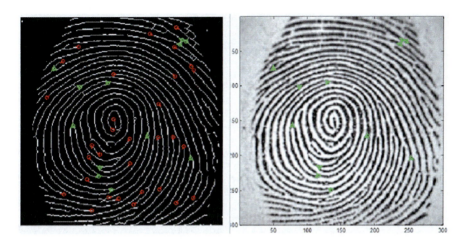

Fig. 17.11 Reckoning of bifurcations

```
BSep=(L==3);
BSepLab=bwlabel(BSep);
propSep=regionprops(BSepLab,'Centroid','Image');
CentSep=round(cat(1,propSep(:).Centroid));
CentSepX=CentSep(:,1);
CentSepY=CentSep(:,2);
plot(CentSepX,CentSepY,'Column')
```

17.2.11 Program Code for Exporting Minutiae Points (Fig. 17.12)

Fig. 17.12 Export of minutiae points

Terminations:

X	Y	Angle
232	16	-0.52
193	25	-0.52
87	43	3.14

Bifurcations:

X	Y	Angle 1	Angle 2	Angle 3
240	34	2.62	-2.09	0.00
246	36	2.62	1.05	-0.52
237	40	3.14	1.05	-0.79

```
function ExportMinutiae(figure, event, userdata)
CentSepX=getappdata(userdata.FPGUI,'CentSepX');
CentSepY=getappdata(userdata.FPGUI,'CentSepY');
OSep=getappdata(userdata.FPGUI,'OSep');
MSep=[CentSepX CentSepY OSep];
CentFinX=getappdata(userdata.FPGUI,'CentFinX');
CentFinY=getappdata(userdata.FPGUI,'CentFinY');
OFin=getappdata(userdata.FPGUI,'OFin');
MFin=[CentFinX CentFinY OFin];
prompt = {'Enter the file name for saving:'};
title = 'Input the values for Minutia export';
num_lines = 1;
def = {'Export Minutiae File'};
answer = input(prompt,title,num_lines,def);
saveMinutiae(answer{1},MFin,MSep);
```

After executing these steps the security of the legitimate user is ensured over the public network because whenever legitimate user would login, exact reckoning value of minutiae points would be considered [8]. And nobody has the same number of minutiae points over the same place of the fingerprint which states the uniqueness of the individual. This will further fortify the trust and privacy of the legitimate user because whenever any malicious user would try to intrude then reckoning value of minutiae points would not match and tracing the malicious user is very easy [9].

17.3 GUI for Evidence of Improved Trust and Privacy of Legitimate User

For proving that RNA-FINNT resulted in the improved trust and privacy for the legitimate user a graphical user interface is designed. This interface is evident for enhancing the security over the public network which is the most insecure network. The command of guide is used in the MATLAB for designing the GUI. Buttons for Binarize [10], Thinning [11], Find Minutiae, Remove False Minutiae, Orientation [12], Validation [13], Reckoning Terminations, Reckoning Bifurcations and Export Minutiae are placed in the GUI (Fig. 17.1a). Explanation of the GUI is as follows:

First of all an RGB image is taken (Fig. 17.1b). This image is three dimensional but in MATLAB forming matrix for 3-D is very difficult so need of conversion from RGB to Grayscale is required. Below is the original RGB image for consideration.

Grayscale image is two dimensional image (Fig. 17.2). Now in MATLAB most convenient mechanism is to work with one dimensional matrix so this originates the need for conversion of grayscale image to binary which would be one dimensional image (Fig. 17.3). Then this Binary image would be thinned. Thinning is one kind of morphological operation (Fig. 17.4) [14]. It is used to remove the pixels from binary image which cause erosion to the image. In case the edges are to be detected in the image then thinning process would be used by reducing all the lines to single thickness of the pixel but also saves the original thickness of the lines. It also reduces the threshold output of an edge. The input is a binary image and the generated output is also a binary image. This procedure of thinning erodes away the boundaries but does not affect the pixels at the end of the lines. So process of thinning is done on the binary image (Fig. 17.4).

Thinned image would be used to find the minutiae points over the fingerprint. Section 17.2.5 focuses upon the code used to find out the exact placing of the minutiae points. While extracting these minutiae points few points could be false also (Fig. 17.5). Example: Termination is the end point of the edge so few ridges could appear as edges and could be treated as terminations. These false terminations could also result in minutiae points. So it will generate a need of removal of these false minutiae points (Fig. 17.6). These local ridge orientations could be removed with various algorithms but in the paper Sect. 17.2.6 states the code for removal of false minutiae points. As the complete fingerprint is too lengthy for reckoning the values of terminations and bifurcations so a region of interest would be generated in which exact values would be considered (Fig. 17.7).

Section 17.2.7 focuses upon the code of deriving this region of interest. It could be done automatically by fixing the coordinate values or manually though cursor values. Already found out minutiae points could be very well observed in this region of interest (Fig. 17.7). After selecting the region of interest orientation of the image is done (Fig. 17.8). Orientation is the angle between x-axis and y-axis. For calculating this angle below mentioned calculation is being done, first slope of line would be calculated then angle between x and y axis.

Slope of Line : $M1 = y1 - cy/x1 - cx$ $M2 = y1 - dy/x1 - dx$
Angle between x and y axis : $\alpha = \tan^{-1} M1 - M2/1 + M1.M2$

After orientation validation of the region of interest is done (Fig. 17.9). This means the values of x and y axis is derived on the basis of minutiae points extracted in Fig. 17.5.

Red color symbolizes the terminations (Fig. 17.10). The values of terminations derived from the considered region of interest are:

Terminations: $X=232, Y=16$ $X=193, Y=25$ $X=87, Y=43$

Green color symbolizes the bifurcations (Fig. 17.11). The values of bifurcations derived from the considered region of interest are:

Bifurcations: $X=240, Y=34$ $X=246, Y=36$ $X=237, Y=40$

Considering x and y coordinates the values for red points over the region of interest are terminations: $X = 229, Y = 44$ $X = 64, Y = 64$ $X = 254, Y = 72$

Considering x and y coordinates the values for red points over the region of interest are bifurcations: $X = 50, Y = 75$ $X = 130, Y = 95$ $X = 88, Y = 101$

After reckoning the terminations and bifurcations complete minutiae points are exported to the text file, which will give the following values:

Name, Date, Number of Terminations and Number of Bifurcations

Text file will generate the above stated values based upon the fingerprint. These values are the evidence of enhanced trust and privacy of the legitimate user as nobody would have the same angle values at the same location (Fig. 17.12). As minutiae points are set of unique points for each individual so legitimate user can very well trust upon the extracted values from their fingerprint.

17.4 Export of Minutiae Points to Database for Ensuring Security in the Public Network

Minutiae Points are unique points on the fingerprint of the legitimate user. Section 17.2 states the process of reckoning the terminations and bifurcations from the fingerprint of legitimate user. The extracted angle values are to be stored in the form of hash code in the database for fortifying the security further. Example (Fig. 17.13): Whenever a session is to be created between the Client and Server there is a need of mutual authentication for proving the legitimacy of each other. Client and Server will authenticate each other. The database at the Server will

17 Reckoning Minutiae Points with RNA-FINNT Augments Trust and Privacy... 227

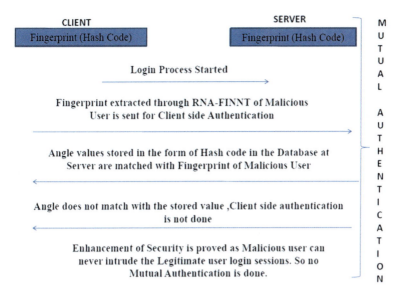

Fig. 17.13 Proof of enhancement of security over the public network

comprise of angle values stored for individual client in the form of hash code. Whenever a malicious user will try to intrude the session between client and server then extraction of their minutiae points from their fingerprint will be done. These extracted values through RNA-FINNT will now be matched with the stored value in the database. And it is but obvious that these values will never match as nobody will have same angle values because of the uniqueness of the minutiae points. This proves the enhancement of security in the public network. Below mentioned is the steps for proving enhancement of security over the public network:

17.5 Future Applications of RNA-FINNT

As RNA-FINNT has resulted in the fortification of trust and privacy for legitimate user and enhancement of security over the public network, below mentioned are the future prospective applications of RNA-FINNT:

1. Online Banking has implemented only password as security feature so fingerprint could be assimilated for enhancement of the security of the legitimate user.
2. Student Attendance System is manually done in many organizations. It could be done with Fingerprint to prove the authenticity and to avoid proxy by the students.
3. Library Management System could be automated by implementing fingerprint of the candidate while issuing and receiving the books. This will enhance the security in the library system.

Future prospects of RNA-FINNT are very bright as it could be implemented for fingerprint authentication in majority of the organizations which work with identity authentication.

References

1. K. Kaur, G. Geetha, Fortification of transport layer security protocol by using password and fingerprint as identity authentication parameters. Int. J. Comput. Appl. **42**(6), 36–42 (2012). Published by FCS, NY, USA
2. V. Goyal, V. Kumar, M. Singh, A. Abraham, S. Sanyal, CompChall: addressing password guessing attacks (2003), http://eprint.iacr.org/2004/136.pdf
3. S. Kapoor, Session hijacking exploiting TCP, UDP and HTTP sessions infosecwriters.com/text_resources/.../SKapoor_SessionHijacking.pdf
4. Hanjae Jeong, Dongho Won, Seungjoo Kim, Weaknesses and improvement of secure hash-based strong-password authentication protocol. Info. Secur. Group J. Info. Sci. Eng. **26**, 1845–1858 (2010)
5. K. Kaur, G. Geetha, Fortification of transport layer security protocol with hashed fingerprint identity parameter. Int. J. Comput. Sci. Issues **9**(2), No.2, 188–193 (2012)
6. A. Farina, Z.M. Kovács-Vajna, A. Leone, Fingerprint minutiae extraction from skeletonized binary images. Pattern Recognit. **32**, 877–889 (1999). Published by Elsevier Science Ltd
7. R. Bansal, P. Sehgal, P. Bedi, Minutiae extraction from fingerprint images – a review. Int. J. Comput. Sci. Issues, **8**(5), No 3, 74–85, (2011)
8. L. Wieclaw, A minutiae based matching algorithms in fingerprint recognition systems. J. Med. Inform. Technol. **13**, (2009). Published by Institute of Informatics, University of Silesia, Będzińska
9. S.D. Patil, S.A. Patil, Fingerprint recognition using minutia matching. World J. Sci. Technol. **2**(4), 178–181 (2012)
10. D. Maltoni, D. Maio, A.K. Jain, S. Prabhakar, *Handbook of fingerprint recognition* (Springer, New York, 2003)
11. A. K. Jain, A. Ross, S. Prabahkar, Fingerprint matching using minutiae and texture features, in *Proceedings of International Conference on Image Processing (ICIP)*, Thessaloniki, 7–10 Oct 2001, pp. 282–285
12. A. Ross, A.K. Jain, J. Reisman, A hybrid fingerprint matcher. Pattern Recognit. **36**(7), 1661–1673 (2003)
13. A.K. Jain, L. Hong, R. Bolle, On-line fingerprint verification. IEEE Trans. Pattern Anal. Mach. Intell. **19**(4), 302–314 (1997)
14. R. Snelick, M. Indovina, J. Yen, A. Mink, Multimodal biometrics: issues in design and testing, in *Proceedings of Fifth International Conference on Multimodal Interfaces*, Vancouver, 2003, pp. 68–72, 1859–1867

Chapter 18
Empirical Study of Email Security Threats and Countermeasures

Dhinaharan Nagamalai, Beatrice Cynthia Dhinakaran, Abdulkadir Ozcan, Ali Okatan, and Jae-Kwang Lee

Abstract Due to wide range of users, email had emerged as one of the preferred method to intrude Local area networks and end users. In this paper we examine the characteristics of various email security threats and the technology used by attackers. In order to counter defense technology, attackers change their mode of operation frequently. The continuous evaluation of attacker's pattern will help the industry to combat attacks effectively. In our study, we collected several thousand spam from a corporate server for a period of 12 months from Jan 2012 to Jan 2013. From the collected data, we identified various types of security threats through email and the attacker's mode of operation. We believe that this study will help to develop more efficient and secure methodologies against security threats through email.

18.1 Introduction

Internet has emerged as a platform for the multibillion dollar illegal business industry. There is no proper economic estimation for the illegal business through the internet. Gartner studies states that the volume of illegal business ranges from 20 to 40 billion dollars. Since the mass adaptation of internet based applications such as cloud computing, social networking, VOIP applications, video streaming and the growing number of mobile users of these applications makes users highly

---

D. Nagamalai (✉) • A. Ozcan • A. Okatan
Department of Computer Engineering, Faculty of Engineering, KTO Karatay University, Konya, Turkey
e-mail: dhinaharann@gmail.com

B.C. Dhinakaran
Optus Networks, Australia, Sydney

J.-K. Lee
Department of Computer Engineering, Hannam University, Daejoen, South Korea

vulnerable to various attacks such as DDoS, phishing and MiM etc., and identity theft.

Due to the rapid growth of Internet, email has emerged as one of the powerful communication tool in modern world. Due to the uncontrolled structure of the internet, email has emerged as a breeding ground for illegal activities from anonymous mail to various attacks such as DDoS attacks, man in the middle attacks, phishing, and several types of scam. The growing number of spam traffic confirms that spamming is one of the profitable businesses for scammers and criminals. Off Late, social media sites are new ways to launch identity theft and spread security threats to end users due to high usage. Security threats are not only emerging as result of software or infrastructure vulnerability as tradition. The user vulnerability plays an important role for security issues in the socially well connected internet era. Email is an easy way to penetrate end users system, to steal data and cripple network services and local network servers.

Criminals launch DDoS, phishing attacks using social engineering techniques. By well crafted email messages the attackers try to infiltrate highly secured local area networks. Spam hosts security threats such as DDoS attacks, phishing, virus, worms and malware attacks, Nigerian fraud, password phishing etc., Some of these spam based security attacks are based on social engineering techniques rather than technological vulnerabilities. Naive users are highly vulnerable to social engineering techniques. Lot of effort had taken to combat spam but there is no mechanism to put an end to spamming activities. Spammers are highly active group of professionals (or criminals) working relentlessly to bypass the anti spamming techniques to reach end users inbox. Spammers use free mail service providers and their own mail servers or services to spread their wings. The arms race between spammer and anti spammers is a never ending race [1]. Due to financial gain of the spamming activity, spammers usually find new ways to spam end users. There are enormous methodologies developed to stop spamming but nothing has stopped the spammers to end their activities. For example spamassasin developed 25 versions of software to combat spam based on spammer's activity. Yet spammers change their mode of operation frequently to escape from anti spam filters or techniques.

Spam consumes server resources, network bandwidth and spread virus, worms, Trojan [2]. Spam is a byproduct of email and causes various security threats like DDoS attacks, Phishing and email scam such as Nigerian fraud mails, password phishing, social security issues [3]. Since online business is growing, new security threats have emerged to gain financial benefits from these thriving industries. These attacks through spam highly depend on social engineering techniques rather than technical innovations.

The remaining sections are organized as follows. Section 18.2 provides background on security threats and various methods used to defend them. Section 18.3 provides data collection and experimental results. In Sects. 18.4 and 18.5 we describe the various threats targeting end users through email. Section 18.6 provides the proposed countermeasure and finally we conclude in Sect. 18.7.

18.2 Related Work

Authors [4] extensively presented cooperative methodology to defend DDoS attacks and internet worms. They simulated experiments using Emulab test bed and evaluated the effectiveness of their proposed cooperative defense mechanism against DDoS attacks and worm threats. In their proposed model, local detection node which can be routers monitors the traffic using existing mechanisms to detect DDoS attacks. If there is any abnormal traffic, it will alert that there will be an attack. When one local node detects the attack pattern, it will pass the message to other nodes using epidemic algorithms. Each local detection node deploys countermeasures to defend DDoS attack in its immediate network. The combination of local detection nodes works to defend DDoS attack in the network. According to authors [5], the hackers use search engines to identifying vulnerable web servers or web sites to host phishing websites as part of their phishing strategy. The authors study reveals that significant numbers (%) of these servers were re compromised after a year. The main reason behind this re-compromise is the security issues are not addressed well by these web servers or websites. By spamming methodology explained by Dhinaharan et al. [3], the phishers use compromised web servers to host phishing web sites and free web site hosting services. The authors analyzed millions of spam mails and identified two types of phishing attacks. According to their work spamming techniques are used to reach user's inbox. Their paper proposed a multi layer approach to defend Phishing attack by a combination of fine tuning of spamming techniques, periodically identifying false negative to spam, blocking IP addresses of the attacker, effective usage of filters, and user education etc., Spam is one of the economical methodology to spread commercials through email, blogs, forums, email archives, social networking sites and messengers [6].

18.3 Network Architecture and Security Setup

Servers providing services to internet users are highly vulnerable than servers located in protected local area networks. The local area network is highly protected by internal firewalls, organized network monitoring and its well separated from the World Wide Web. The servers located in local area network are less prone to major security threats as the ports are closed for external access. Bandwidth consuming applications or systems can be killed in local area network by continues network monitoring. Since the local area network firewall provides another security layer for the backbone any DDoS attack can be thwarted effectively.

18.3.1 Why DDoS Attacks Through Spam?

The easiest way to penetrate into protected Local area network is, infect the end users machine through services offered by Internet. All servers in LAN such as file servers, database servers, CITRIX servers, WLAN authentication servers, backup servers, IP telephone servers, print servers and application servers authenticate users through active directory servers (primary and secondary). Penetrating LAN through WLAN and IPT servers are not useful to access the network. WLAN server and IPT servers that are connected to the World Wide Web are highly vulnerable than other servers located in the LAN. But accesses to these servers are not useful for a DDoS attack. Primary and secondary active directory servers are usually well protected to external threats; the only remaining target is the email server. Attackers can gain easy access to email server by targeting naïve users in the local area network. If the attacker infects one end user's system by social engineering based methodology, he can launch DDoS attacks in the local area network [7]. This scenario can be explained as follows.

Even though there are many layers of security features to provide resilient against security threats and vulnerabilities, yet it has become inevitable. Modern network architectures increasingly deploy additional security systems such as internal firewalls, network monitoring systems, local area network filtering systems apart from the regular security setups as shown in Fig. 18.1. The regular security setups and security services provided by ISPs and government organizations (such as department of education in case of educational institutes) are not enough to meet the requirement of the end users in a network.

The following figure shows the security settings of modern network architecture to protect resources and end users from the attackers.

Internal firewalls and intrusion detection systems deployed in a network are not capable to defend security threats through email services. As security setups are busy protecting server farm and bandwidth exploitation, the attackers are select email services to reach the end users systems safely. Social engineering based security vulnerabilities are capable to defeat the in-built security features available in commonly used operating systems. The operating systems in built-in security features prevent end users system from crashing during attack scenario [8]. Even third party security software installed in end user machines are also highly vulnerable to new type of attacks, virus, worms and Trojans [3]. The attackers use email services within the network and social engineering techniques to launch attacks against network and end users.

18 Empirical Study of Email Security Threats and Countermeasures

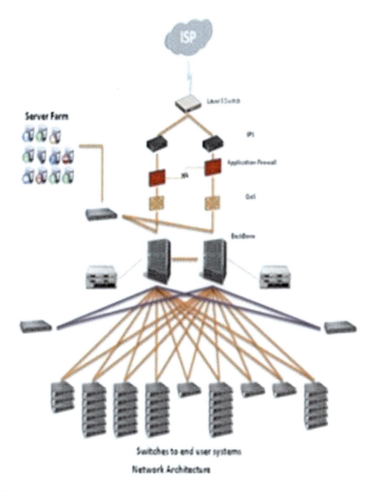

Fig. 18.1 Modern local area network architecture with security setup

18.4 Data Collection

Our corporate mail server is connected to 10 GB backbone and serves more than 300 users. We collected more than 600,000 spam for a period of 12 months from January 2012 to January 2013. The speed of Internet n is 200 Mbps with 60 Mbps upload and download speed. Due to security and privacy concerns, we are unable to disclose the real domain name. From this collection, we separated different category of spam mails by a small program written in Python. In order to counter security mechanisms, attackers change their way of operation frequently. Continuous evaluation of these threats and attacker's technology are required to defend security threats.

From our analysis, we can divide the attacks roughly into two categories such as server based attacks and client based attacks. The DDoS attacks through spam targets servers rather than clients [7]. Phishing attacks, password phishing, Nigerien fraud mails, social network based security threats target client data for financial gain [7, 9]. The client based attacks target end users to collect sensitive data such as banking, information, passwords, etc., to commit financial crimes. But in both cases the attacker is gaining access to end user's data through their inbox using social engineering techniques. Our data collection in the following Table 18.1 shows the number of email security threats for the period of 12 months from Jan 2012 to Jan 2013.

18.4.1 DDoS Attack

DDoS attack still remains one of the major security threats to the networking world starting from service industry to end users [10]. Traditional cloud-based anti-DDoS services are largely ineffective against these application-layer attacks. There is no effective mechanism to stop DDoS attack unless the attacker decides to do so. The Intrusion detection system and firewall offers some security against single user DoS attack but most attacks are distributed and these systems alone are not enough to handle DDoS attack. Continuous DDoS monitoring system will play an important role to defend DDoS attacks.

Regardless of the network infrastructure or bandwidth, DDoS attack can be applied to any network to cripple network services and resources. A report published by Arbor Networks, the World Infrastructure Security Report 2010, and shows that the DDoS attack size has passed 100 Gbps level [11]. All aspects of networking, servers, protocols and services that are vulnerable to DDoS attacks are described by DDoS attack surface. The vulnerable attack surface for DDoS keeps increasing as new equipment, protocols and services are introduced into networks. This presents a significant challenge for network operators. Botnet-driven volumetric and application-layer DDoS attacks continues to be the most significant problem that the operators face.

There are several factor that drive attackers to launch DDoS attack on to the network services and servers. Since tracking the source of DDoS attack is a challenge by the available technology, stopping the attack immediately also has become impossible [7]. The easy availability of sophisticated automated and semi automated tools to launch DDoS attacks are wide spread and this makes the attackers to be anonymous thus leaving law and order in the lurch. Recent DDoS attacks show that major ISPs are highly vulnerable with their huge infrastructure such as 10 GB, 40 GB, 100 GB backbones and server farms [4]. Mostly small and medium size organizations with one or two email servers are highly vulnerable to DDoS attacks launched through spam. As the resources of the servers will be dried up by DDoS attacks through spam, the server will be out of service resulting in

Table 18.1 Email security threats data for 12 months from email server trap

Type of attack	Jan	Feb	March	April	May	June	July	August	Sep	Oct	Nov	Dec
Spam	75,004	70,128	55,986	76,980	80,100	69,876	72,567	58,901	85,615	78,645	76,456	70,980
Virus, worms, Trojon (DDoS)	750	812	656	812	858	870	972	923	867	810	918	1002
Phishing	267	217	201	310	150	310	280	800	279	367	427	189
Nigerian	628	249	486	224	178	240	378	522	320	781	560	1,200

Denial of service [7]. Most of the DDoS attack scenarios shows the major involvement of botnet. In spam based DDoS attacks, the system that is connected in local area network gets converted as botnet during the attack.

18.4.2 DDoS Attack Through Spam

DDoS attack using spam is one of the old version of DDoS attacks. In this type of attack, the infected system becomes a botnet to launch attack on other systems in the same network. The arms race between security service providers and attackers is a never ending problem. In most of attacks, the network has issues after several days from the initial launch of the attack as the resources are eaten up by DDoS attack [7]. This is visible in the case of Sony PlayStation attack and South Korean government web site attacks [12, 13]. The attackers have clearly demonstrated their ability as the target websites started to work after the attacker released some of the web sites from their hit list [12]. There are different types of DDoS attacks but every method tries to exhaust the server's maximum resources such as processing power, memory and band width of the network. The main advantage of DDoS attack is any service can be brought down without exploitation of flaw in the victim service or server. DDoS attack through spam is one of the indirect way of launching attack on a particular target. Mostly this type is used to exploit ignorant end user against local area network of a particular organization rather than software vulnerability. Even though the chances for DDoS attack through spam is less it is not negligible. DDoS attack through spam target email servers to disturb the services rather than the entire network. DDoS attack through spam mail has emerged as one of the common methods to launch DDoS attack on services. Attackers send well crafted email with a small program as an attachment to intrude the local area network. If the user downloads the attachment and opens it, the files attached will be executed and this results in the installation of malware/virus/worm/Trojan in the end users system. Upon execution of the attached file, the server resources will be eaten up by the spearhead emails in the domain machines and this results in the denial of services to legitimate users.

The attackers take maximum effort to pass through the spam filters and deliver the spam mail to the user's inbox. Here the hackers do enough to make the mail recipient to believe that the spam mail is from the legitimate user. The attackers use fake email ids from victim domains to penetrate into the network. The spam mails are usually sent in the name of Network administrator/well wisher of the victim or boss of the organization. Note that the spam mails do not have the signature of the legitimate user. The spam contains small size executable file as an attachment (for example update.exe). The attackers use double file extension to confuse the filter (Update_KB2546_*86.BAK.exe (140 k)) and users. The attachment size ranges from 140 to 180 KB. Mostly the spam mail asks the recipient to execute the .exe file to update antivirus software. Upon execution of the attachment, it will drop new files in windows folder and change the registry file, link to the attacker's website to download big programs to harm the network further. The infected machine collects email addresses from the windows address book and automatically sends mails to others in the same domain. Even if the users don't use mail service programs like Outlook express, it will send mails by using its own SMTP. This type of spam mail attracts group mail ids set up in organizations for easy and time-saving communication. By sending mails to the group, it will spread the attack vigorously. If any of the users forward this mail to others it will worsen the situation. Ultimately the server will receive enormous request from others beyond its processing capacity. In this way it will spread the attack and results in a DDoS attack. After the first mail, for every minute it will send the same kind of mail with different subject name & different contents to the group email ids. Within a day it will eat up server resources and end up in distributed denial of service attack. The names of the worms used in these kind of DDoS attacks are WORM_start.Bt, WORM_STRAT.BG, WORM_STRAT.BR, TROJ_PDROPPER.Q. Upon execution, these worms dropped files namely serv.exe, serv.dll, serv.s, serv.wax, E1.dll, rasaw32t.dll etc.

DDoS malware cause direct and indirect damage by flooding specific targets [10]. Mass mailers and network worms cause indirect damage when they clog mail servers and network bandwidth. This consumes the network bandwidth and resources, causing slow mail delivery further resulting in Denial of service. The server will be down due to enormous request from clients and bulk mail processing.

18.5 Phishing Methodology

Tricking users to divulge their sensitive information such as credit card details, login, passphrase etc., is referred as Phishing [5]. Phishers accomplish this task by sending well designed spam mail with attachments that contains virus, worms and Trojans which the end user will accidently install on their machine. Recently phishing attacks are emerged from social networking sites. The different types of phishing methods are explained in the following sections.

Method 1: Phishers send email message to end users pretending it is from a financial institution and ask them to reveal their personal details by clicking the link given in the mail [3, 14]. By clicking the link, the user will be taken to fraudulent web page which is controlled by the Phisher. The fraudulent web page is designed exactly similar to the original one and therefore the user never suspects there is a foul play. The data entered in the fake website by the end user will be collected and misused by the phishers and thereby leading to data and financial loss. There will also be a reputation loss for the victim's institution. Phishers use botnet, small business based email services, free email services to spam end users in this category.

Method 2: Phishers convince end users to download attachment from email. The attachment will contain malware, virus, worms and Trojans. Upon execution of the downloaded attachment, the files will be dropped to the end users system and

Table 18.2 List of virus, worm and Trojan

Name of virus, worm, Trojan
Spam.Phish.url
Spam.Hoax.HOAX_PHISH_FORGED_PAYPAL
Clm.HTML.Phishing.Pay-110
Spam.Hoax.HOAX_PHISH_FORGED_EBAY
Spam.Hoax.HOAX_PHISH_FORGED_CITIBNK
Trojan.spy.html.bankfraud.od

this alters the registry settings of the system, installs spyware or remotely controls client system. By using these files, the phisher can collect sensitive information such as passwords, credit cards, bank information etc.,. There are different kinds of malware, worms, virus designed to target a particular brand or users. For example spam.hoax_phish_forged_citibank targets Citibank ATM, Debit card and pin card information [3]. The following table shows various types of virus, worms and Trojans (Table 18.2).

There are hundreds of variants for each of these viruses, worms and malware. Based on the defense mechanism devised by the anti spam technology, each version gets updated.

18.6 Email Phishing

Email phishing is a byproduct of phishing attacks to steal end users sensitive information by hacking email accounts by social engineering techniques. In phishing, criminals try to get sensitive information for financial gain. But in the case of email phishing, the hackers steal the email account, to commit various frauds by impersonating to be the authentic real user. Examples are sending emails saying that someone in the family is very sick and needs money for a speedy recovery or email posing as the service provider or network administrator by requesting the end users to divulge their account details to avoid shutting down of the account or upgrading the inbox capacity and so on., If the end user sends the details to the phisher, the email account control will be taken over by them. The hackers use phished email account for several criminal activities such as stealing personal information from the inbox, sending mail to other contacts in address book, spamming others. If the hacker gets banking information, credit card details and other sensitive credentials, it will lead to identity theft and financial loss. In some cases, money request email will be sent to all email accounts saved on the address book saying that the "real end user" is stuck somewhere in a foreign land and needs money to get back home [3, 15]. If free email service providers start to tighten their rules to create new email accounts, then old email accounts will help the spammer to spam end users.

18.6.1 *Spearhead Phishing*

After downloading and execution the attachment, the worm, virus and Trojan installs or drop files in the target system. The newly installed malware, virus and Trojan will take control of the email address book of the end users. The virus will send phishing mails or attack all email accounts in the address book. This method is referred as spear head phishing. The working principle of spearhead phishing is similar to the methodology of DDoS attacks created by Spam [7]. These attackers take additional efforts to reach the end users system by attachments with double extensions to bypass filters [7].

18.7 Nigerian Scam

Nigerian fraud mail is one of the successful letter frauds over the Internet [9]. Fraud is one of the oldest ways of making living and used to be very popular in the western world. As long as there are vulnerable people in the world, there will be a fraud. Off late, emerging economies like India, Brazil etc. are being targeted. Initially fraud mails originated from Nigeria but now days it comes from all over Africa, European Union and Asia. Due to anonymous nature of the email, it is very hard to tackle this type of fraud. Nigerian fraud mail is not based on any technical innovations. Fraudster does not rely on any technical skill but just have the art of convincing a willing victim to responding to the invitation and to send money in several installments for a variety of reasons to sender of the mail. Nigerian fraud is clearly based on social engineering techniques to lure end users to share in a percentage of millions of dollars that the sender—a self-proclaimed government official—is trying to transfer illegally out of Nigeria. Based on social engineering techniques, Nigerian fraudsters lure "greedy, easy money makers" to their trap. Small business, religious organization's authorities, and nonprofit organizations are primarily targeted. There are many variations on the way the scam works such as "lottery prize", "transfer of funds", "chemicals to transform papers to USD bills" each naming different persons and describing different circumstances.

In simple words, Nigerian fraud works by blinding the email users with promises of unimaginable wealth transfer from an African nation or a "troubled nation or a person". To do this transaction, the end users are asked to pay significant money to the Nigerian mail sender. In reality, the millions of dollars do not exist, and the victim eventually ends up with nothing but loss. Once the victim stops sending money, the fraudsters use the personal information and checks that they received to impersonate the victim, draining bank accounts and credit card balances.

```
  From: Colin Moulder                                                                 Tuesday, July 12, 2011 10:44 PM
  From: "COLIN MOULDER" <clm1@bk.ru>
  To: undisclosed-recipients

Dear friend,
How are you and your family? I know my letter meets you in your best mood today; my name is Mr. Colin Moulder,  I decided to seek a
confidential co-operation with you in the execution of the deal described here-under for our both mutual benefit and hope you will keep it a
top secret because of the nature of the transaction, While  auditing  the bank accounts, I discovered an unclaimed /abandoned fund, sum
total of 23.800,000 GBP (Twenty Three Million Eight Hundred Thousand British Pounds Sterling) in an account that belongs to one of our
foreign customers an American who unfortunately lost his life with his wife, in a plane crash in September 11, 2001.
My request for foreigner to stand as next of kin in this business is because the fact that the customer is a foreigner. Based on the fact
that this is a deal, I propose that 25% of this fund is for you, and 75% is for me. I will give more detail of the transfer process as soon
as you show your interest in this transaction.
Please note that there will be no problem as my bank has made all effort through to reach for any of his relation but all was fruitless. My
position as the chief auditor in this bank guarantees the successful execution of this (deal) transaction. This money will be shared on
agreed percentage which must be written in the agreement of personal trusteeship, once the fund is transferred to your account in your
country or any country of your choice.
Finally, If not interested please destroy the mail because of the confidentiality attached to it. I promise that this transaction will be
done under a legitimate arrangement that will protect both of us from any breach of the law. Please feel free to contact me immediately at
(+44-703-590-4109). Upon consideration and acceptance of this offer, please send the following information to me immediately. 1. -Your Full
Name, 2 -Your Contact Address 3-Your direct Mobile telephone Number.
Thanks.
Mr. Colin Moulder
Private Email:clmoulder1900@mail.kz
```

These days due to increase in the number of mobile phone users, Nigerian fraudsters send bulk SMS to attract end users. After initial SMS contact, the fraudsters use email contact to dupe the users. The scam email describes that user's mobile numbers is selected under online lottery system and has won huge amount of money (for example 1 billion dollars). If the end user contacts the fraudster, they send step by step procedure through email to collect so called "lottery prize". The step by step procedure to collect lottery money will lead to scam as we described earlier. The fraudsters use different proxy bank accounts to collect initial fee from victims. As long as the victim does not raise alarming issues, the fraudster will be in touch with the victim. According to Wendy et al. [9], Nigerian fraud mail is one of the major fraudulent activity against North American end users. Even though the Nigerian fraudsters are spread across the world, the North American region has the more active scammers [9]. Since these types of scams are new to developing nations like India, Brazil and China the user awareness is lesser than developed nations and thereby the number of victims are more. There is no security threat to either the network or systems. Though the majority of the email users ignore these types of scam, there people who still believe the well crafted scam email.

18.8 Countermeasures

There are many antivirus and anti spamming software available to combat security threats yet email is one of the major source for security threats. Attackers always come up with new mechanism to defeat security settings [2]. Email security threats are heavily dependent on social engineering techniques rather than technological vulnerabilities. Spamming, DDoS attacks, Phishing and Nigerian scam are by products of email system [2]. Effective anti spam techniques should be implemented to defend security threats through email. In addition to all security settings, user awareness plays an important role to combat threats. End users should be educated by the organization to overcome these security threats. Email policy should be implemented without any compromise [3]. Frequently analyzing email security threats helps to better understand the pattern of attacker's behavior. Failure to maintain an updated understanding of these threats will leave end users highly vulnerable to security threats through email.

18.9 Conclusion

As email has emerged as the most favored medium of communication, the security threats through email has become common and easy. Our study shows that email has emerged as a launching pad for security threats such as DDoS, Phishing, MiM and Nigerian scam. The uncontrolled nature of email services makes it is easier to launch attacks on end users. Hackers use spam mails to launch DDoS attack on corporate email servers to cripple their network. Monitoring the network continuously makes launching of DDoS attack harder but we cannot completely eliminate the risk. By using social engineering techniques in Phishing, hackers try to steal sensitive information from users to commit financial crimes. Nigerian scams are aimed to steal money from individuals by well crafted emails to transfer money. In the case of email phishing, hackers try to steal the email account to commit other frauds. Hackers use the stolen email accounts to gain information such as bank details, credentials of other secured systems etc. Even though there are several software tools to stop these criminal activities, there is no bullet proof mechanism to completely eliminate such activities. User awareness plays a vital role to defend these threats effectively. Continues evaluation of these threats will help the industry to design new ways to handle these threats effectively. Based on our study, we have designed a security tool to combat email security threats effectively and it will be tabled in our future work.

References

1. B. Cynthia Dhinakaran, Jae-Kwang Lee, D. Nagamalai, An empirical study of spam and spam vulnerable email accounts, in *Proceedings of FGCN-2007*, IEEE CS Publications, vol. 1, Dec 2007, pp. 408–413
2. B. Cynthia Dhinakaran, Jae-Kwang Lee, D. Nagamalai, Characterizing spam traffic and spammers, in *Proceedings of ICCIT 2007*, IEEE CS Publications, Nov 2007, pp. 831–836
3. C. Dhinakaran, D. Nagamalai, Jae-Kwang Lee, Multilayer approach to defend phishing attacks. J. Internet Technol. **11**(3), pp. 417–426 (2010)
4. G. Zhang, M. Parashar, Cooperative detection and protection against network attacks using decentralized information sharing [last]. Cluster Comput. **13**(1), 67–86 (2010)
5. T. Moore, R. Clayton, The impact of public information on phishing attack and defense. Commun. Strateg. **81**, 47–67 (2011)
6. B. Dhinakaran, D. Nagamalai, Jae-Kwang Lee, Bayesian approach based comment spam defending tool. Adv. Info. Secur. Assur. LNCS **5576**, pp. 578–587 (2009)
7. D. Nagamalai, B. Cynthia Dhinakaran, Jae-Kwang Lee, Multi layer approach to defend DDoS attacks caused by spam, in *Proceedings of IEEE MUE 07*, April 2007, pp. 97–102

8. S. Kumar, S. Surisetty, Microsoft vs Apple: resilience against distributed denial of service attacks, *IEEE Security & Privacy*, March–April 2012, pp. 60–64
9. W.L. Cukier, E.J. Nesselroth, S. Cody, Genre, narrative and the "Nigerian Letter" in electronic mail, in *Proceedings of 40th HICSS-2007*, pp. 1–10
10. J. Francois, I. Aib, R. Boutaba, Firecol: a collaborative protection network from the detection of flooding DDoS attacks. IEEE/ACM Trans. Netw. **20**(6), 1828–1841 (2012)
11. Arbor networks' sixth annual worldwide infrastructure security report, Reveals DDoS attack size breaks 100 Gbps for first time; Up 1000 % Since 2005, 1 Feb 2011, http://www.allvoices.com/news/8061300-arbor-networks-sixth-annual-worldwide-infrastructure-security-report-reveals-ddos-attack-size-breaks-100-gbps-for-first-time-up-1000-since-2005
12. B. Harden, South Korean web sites are hobbled in New Round of attacks, The Washington Post, 10 July 2009, http://articles.washingtonpost.com/2009-07-10/world/36838896_1_web-sites-computer-security-intelligence-agency
13. Sony tells congress anonymous DDoS aided breach, Infosec island, 5 May 2011), http://www.infosecisland.com/blogview/13558-Sony-Tells-Congress-Anonymous-DDoS-Aided-Breach.html
14. T.F. Stafford, R. Poston, Online security threats and Computer User Intentions, IEEE CS Computers Magazine, pp 58–63 (2010)
15. C. Dhinakaran, Jae Kwang Lee, D. Dhinaharan Nagamalai, Reminder: please update your details: phishing trends (2009, December). IEEE Xplore, in *Networks and Communications, 2009. NETCOM'09*. pp. 295–300

Chapter 19
Improving Business Intelligence Based on Frequent Itemsets Using k-Means Clustering Algorithm

Prabhu Paulraj and Anbazhagan Neelamegam

Abstract In this world, each and every activity is enriched with lot of information. Business and other organization needs information for better decision making. Business Intelligence is a set of methods, process and technologies that transform raw data into meaningful and useful information. Some of the functions of business intelligence technologies are reporting, Online Analytical Processing, Online Transaction processing, data mining, process mining, complex event processing, business performance management, benchmarking and text mining. The applications of business intelligence includes E-commerce recommender system, approval of bank loan, credit/debit card fraud detection etc., In order to obtain business intelligence from large dataset there many techniques are available in data mining such as characterization, discrimination, frequent itemset mining, outlier analysis, cluster analysis and so on. In this proposed algorithm frequent itemset mining and clustering algorithm is used to extract the information from the dataset in order to make the decision making process more efficient and to improve the business intelligence.

19.1 Introduction

Data Base Management System (DBMS) and Data Mining (DM) are two emerging technologies in this information world. Knowledge is obtained through the collection of information. Information is enriched in today's business world. In order to maintain the information, a new systematic way has been used such as database.

P. Paulraj (✉)
Department of Information Technology, DDE, Alagappa University,
Karaikudi, Tamil Nadu, India
e-mail: pprabhu70@gmail.com

A. Neelamegam
Department of Mathematics, Alagappa University, Karaikudi, Tamil Nadu, India
e-mail: anbazhagan_n@yahoo.co.in

In this database, there are collection of data organized in the form of tuples and attributes. In order to obtain knowledge from a collection of data, business intelligence methods are used. Data Mining is the powerful new technology with great potential that help the business environments to focus on only the essential information in their data warehouse. Using the data mining technology, it is easy for decision making by improving the business intelligence.

Frequent itemset Mining is to find all the frequent itemsets that satisfy the minimum support and confidence threshold. Support and Confidence are two measures used to find the interesting frequent itemsets. In this paper, frequent itemset mining can be used to search for frequent item set in the data warehouse. Based on the result, these frequent itemsets are grouped into clusters to identify the similarity of objects.

Cluster Analysis is an effective method of analyzing and finding useful information in terms of grouping of objects from large amount of data. To group the data into clusters, many algorithms have been proposed such as k-means algorithm, Fuzzy C means, Evolutionary Algorithm and EM Method. These clustering algorithms groups the data into classes or clusters so that object within a cluster exhibit same similarity and dissimilar to other clusters. Thus based on the similarity and dissimilarity, the objects are grouped into clusters.

In this paper, FIk-means algorithm is proposed to improve the business intelligence based on frequent itemsets using k-means clustering. This algorithm is experimentally tested with E-commerce application for better decision making by recommending top n products to the customers.

19.2 Related Work

Alexandre et al. [1] presented a framework for mining association rules from transactions consisting of categorical items where the data has been randomized to preserve privacy of individual transactions. They analyzed the nature of privacy breaches and proposed a class of randomization operators that are much more effective than uniform randomization in limiting the breaches.

Jiaqi Wang et al. [4] stated that Support vector machines (SVM) have been applied to build classifiers, which can help users make well-informed business decisions. The paper speeds up the response of SVM classifiers by reducing the number of support vectors. It was done by the K-means SVM (KMSVM) algorithm proposed in the paper. The KMSVM algorithm combines the K-means clustering technique with SVM and requires one more input parameter to be determined: the number of clusters.

M. H. Marghny et al. [5], stated that Clustering analysis plays an important role in scientific research and commercial application. In the article, they proposed a technique to handle large scale data, which can select initial clustering center purposefully using Genetic algorithms (GAs), reduce the sensitivity to isolated point, avoid dissevering big cluster, and overcome deflexion of data in some degree that caused by the disproportion in data partitioning owing to adoption of multi-sampling.

Wenbin Fang et al. [11], presented two efficient Apriori implementations of Frequent Itemset Mining (FIM) that utilize new-generation graphics processing units (GPUs). The implementations take advantage of the GPU's massively multi-threaded SIMD (Single Instruction, Multiple Data) architecture. Both implementations employ a bitmap data structure to exploit the GPU's SIMD parallelism and to accelerate the frequency counting operation.

Ravindra Jain [8], explained that data clustering was a process of arranging similar data into groups. A clustering algorithm partitions a data set into several groups such that the similarity within a group was better than among groups. In the paper a hybrid clustering algorithm based on K-mean and K-harmonic mean (KHM) was described. The result obtained from proposed hybrid algorithm was much better than the traditional K-mean and KHM algorithm.

David et al. [3], described a clustering method for unsupervised classification of objects in large data sets. The new methodology combines the mixture likelihood approach with a sampling and sub sampling strategy in order to cluster large data sets efficiently. The method was quick and reliable and produces classifications comparable to previous work on these data using supervised clustering.

Risto Vaarandi [9], stated that event logs contained vast amounts of data that can easily overwhelm a human. Therefore, mining patterns from event logs was an important system management task. The paper presented a novel clustering algorithm for log file data sets which helps one to detect frequent patterns from log files, to build log file profiles, and to identify anomalous log file lines.

R. Venu Babu and K. Srinivas [10] presented the literature survey on cluster based collaborative filter and an approach to construct it. In modern E-Commerce it is not easy for customers to find the best suitable goods of their interest as more and more information is placed on line (like movies, audios, books, documents etc.). So in order to provide most suitable information of high value to customers of an e-commerce business system, a customized recommender system is required. Collaborative Filtering has become a popular technique for reducing this information overload. While traditional collaborative filtering systems have been a substantial success, there are several problems that researchers and commercial applications have identified: the early rater problem, the sparsity problem, and the scalability problem.

B. Sarwar et al. [9] analyze difference item-based recommendation generation algorithms. Took different techniques for computing item-item similarities and different techniques for obtaining recommendations from them (e.g., weighted sum vs. regression model). Experimentally evaluated their results and compared them to the basic nearest neighbour approach.

P. Prabhu et al. [10] introduced method for determining the optimum number of clusters in a partition simply by examining various cluster validity measures for different values of numbers of clusters. This method finds optimum number of cluster efficiently.

P. Prabhu et al. [11] proposed method for Improving the Performance of K-Means Clustering for High Dimensional Data Set. This experiment shows improvement in accuracy of the clustering results by reducing the dimension and improved initial centroid by partitioning projected dataset and finding the median as centroids.

Zan Huang et al. [12] proposed goal is to develop a metalevel guideline that "recommends" an appropriate recommendation algorithm for a given application demonstrating certain data characteristics.

19.3 Proposed Method

In this business world, there exists a lot of information. It is necessary to maintain the information for decision making in business environment. The decision making consists of two kinds of data such as OnLine Analytical Processing (OLAP) and OnLine Transactional Processing (OLTP). The former contains historical data about the business from the beginning itself and the later contains only day-to-day transactions on business. Based on these two kinds of data, decision making process can be carried out in out by means of frequent itemsets mining and clustering using k-means algorithm in order to improve the business intelligence. The steps involved in the proposed methodology FIk-means algorithm is given below:

- Identifying the dataset
- Choose the consideration columns/features
- Define the rules for identifying frequent itemsets
- Generate the resultant dataset based on rules
- Clustering objects using k-means clustering

19.3.1 Identifying the Dataset

To maintain the data systematically and efficiently, database and data warehouse technologies are used. The data warehouse not only deals with the business activities but also contains the information about the customer that deals with the business. The representation of the data set is shown below:

$$D = \Sigma (A) = \{a_1, a_2, \ldots\ldots\ldots, a_n\} \qquad (19.1)$$

Where, $\Sigma (A)$ is the collection of all attributes, a_1, a_2, are the attribute list that deals with the dataset. Upon collecting the data, the dataset contains the data as follows:

$$\begin{pmatrix} a_{00} & a_{01} & a_{02} & \ldots \ldots \ldots & a_{0n} \\ a_{10} & a_{11} & a_{12} & \ldots \ldots \ldots & a_{1n} \\ & \vdots & & & \\ a_{m0} & a_{m1} & a_{m2} & \ldots \ldots \ldots & a_{mn} \end{pmatrix}$$

Here, a_{ij} is the data elements in the dataset, where $i = 0,1,\ldots n$ and $j = 0,1,\ldots m$.

19.3.2 Choosing the Considering Column/Features

Upon the dataset has been identified, the next step of the proposed work is to choose the consideration column or filtering columns/features. That is, from the whole dataset, the columns/subset of features to be considered for our work has been chosen. This includes the elimination of the irrelevant column in the dataset. The irrelevant column/feature may the one which provide less information about the dataset.

$$CC = \Sigma\left(A'\right) = \Sigma(A) - \Sigma(A_u) \quad (19.2)$$

$$\Sigma\left(A'\right) = \{a_1, a_2, \ldots, a_n\} - \{a_{u1}, a_{u2}, \ldots, a_{un}\} \quad (19.3)$$

$$CC = D' = \Sigma\left(A'\right) = \{a_1', a_2', \ldots, a_n'\} \quad (19.4)$$

Where, CC denotes the consideration column, which will be represented as $\Sigma(A')$, $\Sigma(A)$ represents the set of all attributes in the chosen dataset, $\Sigma(A_u)$ represents the set of all features to be eliminated to get the subset of features. The consideration column of the dataset can be represented as follows:

$$\begin{pmatrix} a'_{00} & a'_{01} & a'_{02} & \ldots \ldots \ldots & a'_{0n} \\ a'_{10} & a'_{11} & a'_{12} & \ldots \ldots \ldots & a'_{1n} \\ & \vdots & & & \\ a'_{m0} & a'_{m1} & a'_{m2} & \ldots \ldots \ldots & a'_{mn} \end{pmatrix}$$

Here, a'_{ij} is the data elements in the new resultant dataset with consideration column, where $i = 0, 1, \ldots, n$ and $j = 0,1,\ldots m$.

$$S = f(X)/D \quad (19.5)$$

This will identify the column S from the dataset.

19.3.3 Define the Rules

From the consideration dataset, the objects can be grouped under stated conditions that are defined in terms of rules. That is, for each column that is considered, specify the rule to extract the necessary domain from the original dataset. This rule is considered to be the threshold value. The domain can be chosen by identifying the frequent items from the dataset. The frequent items can be identified by analyzing the repeated value in the consideration column. Rule can be defined as;

$$FIS = \text{value}(S) > (SUP(S) \text{ and/or } CONF(S)) \quad (19.6)$$

Where, FIS represents the identified frequent itemset. Value(S) is the frequent items in the column S, satisfying the SUP(S) and CONF(S). SUP(S) is defined as the percentage of objects in the dataset which contain the item set. CONF(S) is defined as SUP(X U Y)/SUP (X). (i.e.,) the confidence on the frequent item set can be determined by combining the X and Y values from the dataset and then neglecting the X value to obtain the frequent item.

19.3.4 Generate the Resultant Dataset Based on Rules

The next step upon defining the rules for identifying frequent items is to execute the rules on the new dataset formed by the consideration column. Any objects that satisfy the criteria are selected and counted. This can be carried out by:

$$C_n = \eta \left(a_{ij}(A)/D' \right) \quad (19.7)$$

It counts the number of domains, a_{ij} of the attribute list, A from the Dataset, D. From the counted value, C_n, we can determine the frequent item set that has been occurred in the dataset using the threshold value T.

$$C_n(a_{ij}(A)) > T \quad (19.8)$$

It shows that the domain a_{ij} of attribute A satisfies the threshold value T specified and hence the item a_{ij} is considered to be frequent item set. The identified frequent item sets from the dataset D' are listed as below:

$$\begin{pmatrix} a_{01} & c_1 \\ a_{02} & c_2 \\ \vdots & \\ a_{mn} & c_n \end{pmatrix}$$

Where, a_{01}, a_{02}, a_{mn} are the domain list that occurred frequently in the attribute and

Fig. 19.1 Process involved in proposed method

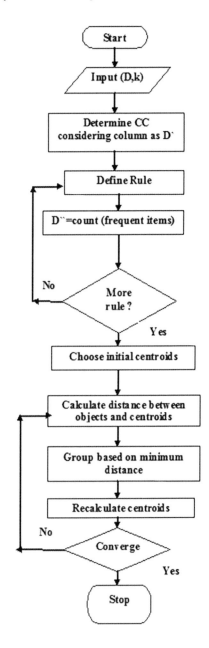

meeting the threshold value specified. c_1, c_2, c_n specifies the number of times that the domain a_{ij} occur in the dataset. This resultant dataset D'' is used to cluster the objects using k-means clustering.

19.3.5 Clustering Objects Using k-Means Clustering

Upon forming the new dataset D″, the objects in D″ are clustered based on similarity of objects using k-means clustering.

k-means clustering is a method of classifying or grouping objects into k clusters (where k is the number of clusters). The clustering is performed by minimizing the sum of squared distances between the objects and the corresponding centroid. The resultant consists of cluster of objects with their labels/classes.

19.3.6 Process Involved in Proposed Method

The Fig. 19.1 shows the flowchart of process involved in proposed Method.

19.3.7 FIk-Means Algorithm

```
FIk-means Algorithm (D, k)
        Input: The number of clusters k, dataset D with n objects.
        Output: A set of clusters Cₖ.
Begin
        Identify the dataset D = Σ (A) = {a₁, a₂, ............, aₙ} attri-
        butes/objects.
        Outline the Consideration Column(CC) from D.
        CC = D' = Σ (A') = {a₁', a₂', a₃' ........ aₘ'}
        Initialize j =0
        Repeat
          Formulate the rules for identifying the similar objects.
          S = f(X)/D, where S is the sample set containing identified
          column
          FIS = value(S) > (SUP(X) and/or (SUP(X U Y)/SUP (X))),
          where FIS is the frequent itemsets identified.
          Cₙ(aᵢⱼ(A')) > T from FIS, where T specifies the threshold value.
          Generate the Resultant Dataset, D″ from Cₙ
          j++
        Until no further partition is possible in CC.
        Identify the k initial mean vectors (centroids) from the objects
        of D″.
        Repeat
            Compute the distance, β between the object aᵢ and the
            centroids cⱼ.
```

$$\beta = \sum_{j=1}^{k}\sum_{i=1}^{n}\sum |a_i^{(j)} - c_j|^2$$

```
Where β Euclidean distance,
        cj the cluster centers/centroids,
        ai objects.
    Choose min{∂(cjk)} as the refined centroids by means of β
    Assign objects to cluster with min{∂(cjk)}
    Recalculate the k new centroids cj from the new cluster formed
Until reaching convergence
End
```

19.4 Experimental Setup

The proposed methodology FIk-means algorithm provides solution to improve the business intelligence. This methodology can be verified through various experimental setups. In this work E-Commerce dataset is used for testing the proposed algorithm for recommending products purchase by the customers. This algorithm help customers find items they want to buy from a business. The E-commerce business, using recommender systems are Amazon.com, CDNOW.com, Drugstore.com, eBay, MovieFinder.com and Reel.com. The Table 19.1 shows the description of sample dataset.

19.5 Results and Discussion

Based on the identified frequent item set mining, it is clear that the customer with age between 21 and 25, frequently accessing the site. And so, the frequent item set being the customer with age less than 25 and greater than 21. From the original dataset, proposed method identified the products purchased by the customers belonging to these age group is shown in Table 19.2.

Table 19.1 Dataset description

Key element	Description
Dataset name	E-commerce dataset (synthetic/realworld)
Original attribute-list	Name, age, gender, occupation, salary, date, time, item purchased, amount
Consideration column	Age, gender, occupation, salary, amount
Rules defined	Customer having age with 21, 22, 23, 24, 25
	Occupation = "Teacher"
	Amount > 25,000

Table 19.2 Identified products purchased by customer

Age	Item codes	Frequently purchased products
21	1010	Computer
22	2001	Jewels
23	3001	Books
24	5010	Sports
25	4101	Shoes

Table 19.3 Initial centroids

Cluster	Age	Mean vector (centroid)
1	21	(1, 1)
2	23	(5, 7)

Table 19.4 Sample iterations of clustering objects k = 2

	Cluster-1		Cluster-2	
Step	Age	Mean vector (centroid)	Age	Mean vector (centroid)
1	21	(1, 1)	23	(5, 7)
2	21, 22	(2, 2.5)	23, 24	(5.5, 4.5)
3	21, 22	(2, 2.5)	23, 24, 25	(5, 4)

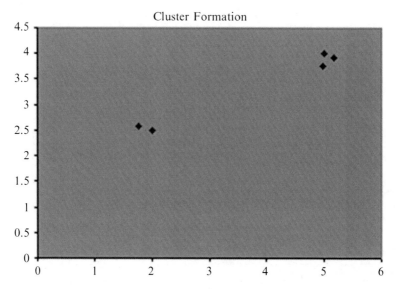

Fig. 19.2 Clusters identified for k = 2

The identified frequent items based on the defined rules are counted to form the resultant dataset D″. Now k-means clustering is applied to group the objects based on similarities. The Table 19.3 shows the initial centroids.

The Table 19.4 shows the sample iterations for clustering objects for k = 2.

Table 19.5 The distance between the objects to the centroid of each cluster

Age	Distance between the object to the centroid in cluster-1	Distance between the object to the centroid in cluster-2
21	1.8	5
22	1.8	2
23	5.4	3
24	4.0	2.2
25	2.1	1.4

The resulting objects found in the cluster-1 are 21 and 21. The objects found in the cluster-2 rare 23, 24 and 25. The Fig. 19.2 shows the clusters identified using the proposed method for k = 2.

The Table 19.5 shows the distance between the object to their centroid of each clusters.

Each individual's distance to its own cluster mean should be smaller that the distance to the other cluster's mean. Thus the mean of the object 21, 22 belongs to cluster-1 which is nearer to cluster-1, whereas the mean of the object 23, 24, 25 belongs to cluster-2, which is nearer to cluster-2. Thus there is no relocation occur in this example. From the obtained results, it is clear that the customers are grouped under 2 clusters named as cluster-1 and cluster-2. Thus, it is easy to identify what are the products that can be recommeded to the customers when they accessing the site.

When the customer enters into the site, the first step is to verify the customer details to identify, on which cluster the customer can be fall on. This can be done by analyzing the age of the customer. Based on the age, the customer can be easily classified and identified their position on the clusters. In our experiment, if the customer belongs to age 22, then they fall under cluster-1 and then we conclude that this customer can have more probability to purchase either computer or jewels. Thus the upcoming customer can now be recommended and redirected to the web page containing the details of computers and jewels. If suppose, the customer belongs to the age 24 means, they fall under cluster-2 and they can have more probability to purchase the products books, shoes or sports items. So they are recommeded to puchase those products by redirecting to that web page. Through this kind of redirection, it is easy for the customer to save the time to search for their desired product. Thus, by grouping the similar behavior customers into a cluster and based on the cluster result, the customer can be recommended and redirected to that products web page. Hence this intelligence provides customers/business for better decision making by recommending the top products.

Also, we can compare the performance of our proposed algorithm with the existing methodologies like [10] with various metrics like precision, recall and silhoutte index. This can be carried out by finding quality with various number of neighbours, iterations and clusters. The performance of the proposed method performs better than existing methods.

19.6 Conclusion and Future Work

Business intelligence is a new technology for extracting information for a business from its user databases. In this paper we presented and evaluated algorithm for improving business intelligence for better decision making by recommeding products purchase by the customer. The performance of the methodology can be verified by undertaking many experimental setups. The results obtained from the experiments shows that the methodology performs well. This algorithm can be tested with many real-world datasets with different metrics as a future work.

References

1. A. Evfimievski, R. Srikant, R. Agrawal, J. Gehrke, Privacy preserving mining of association rules, IBM Almaden Research Center, USA, Copyright 2002
2. B. Sarwar, G. Karypis, J. Konstan, J. Riedl, Item-based collaborative filtering recommendation algorithms. in *Procedings of the 10th WWW International Conference*, Hong Kong, ACM 1-58113-348-0/01/0005, 1–5 May 2001, pp. 285–295
3. D.M. Rocke, Jian Dai, Sampling and subsampling for cluster analysis in data mining: with applications to sky survey data. Data Mining Knowl. Discov. **7**, 215–232 (2003)
4. Jiaqi Wang, Xindong Wu, Chengqi Zhang, Support vector machines based on K-means clustering for real-time business intelligence systems. Int. J. Bus. Intell. Data Mining **1**(1), 54–64 (2005)
5. M.H. Marghny, R.M. Abd El-Aziz, A.I. Taloba, An effective evolutionary clustering algorithm: Hepatitis C case study, Computer Science Department, Egypt. Int. J. Comput. Appl. (0975-8887) **34**(6), 1–6 (2011)
6. P. Prabhu, Method for determining optimum number of clusters for clustering gene expression cancer dataset. Int. J. Adv. Res. Comput. Sci. **2**(4), 315 (2011)
7. P. Prabhu, N. Anbazhagan, Improving the performance of k-means clustering for high dimensional dataset. Int. J. Comput. Sci. Eng. **3**(6), 2317–2322 (2011)
8. R. Jain, A hybrid clustering algorithm for data mining. CCSEA, SEA, CLOUD, DKMP, CS & IT 05, pp. 387–393 (2012)
9. R. Vaarandi, A data clustering algorithm for mining patterns from event logs. in *Proceedings IEEE Workshop on IP Operations and Management*, Kansas City, Missouri, 2003
10. R. Venu Babu, K. Srinivas, A new approach for cluster based collaborative filters. Int. J. Eng. Sci. Technol. **2**(11), 6585–6592 (2010)
11. Wenbin Fang, Mian Lu, Xiangye Xiao, Bingsheng He, Qiong Luo, Frequent itemset mining on graphics processors, in *Proceedings of the Fifth International Workshop on Data Management on New Hardware*, Providence, Rhode-Island, 2009
12. Zan Huang, Daniel Zeng, Hsinchun Chen, A comparison of collaborative-filtering recommendation algorithms for E-commerce. IEEE Int. Syst. **22**(5), 68–78 (2007), IEEE Computer Society

Chapter 20
Reputation-Based Trust Management for Distributed Spectrum Sensing

Seamus Mc Gonigle, Qian Wang, Meng Wang, Adam Taylor, and Eamonn O. Nuallain

Abstract One of the solutions to the hidden node problem in Cognitive Radio (CR) networks is to construct a global radio environment map (REM) hosted on a central controlling server. With this approach, the responsibility of sensing the radio environment can be distributed among all the nodes in the cognitive radio network. This introduces vulnerability because the server depends on the participating nodes to provide honest and accurate spectrum sense information. This research develops a reputation-based security mechanism that protects the radio environment map against falsified spectrum information that may be provided by malicious members of the network.

20.1 Introduction to Cognitive Radio and Radio Environment Mapping (REM)

In traditional methods of radio frequency allocation, bands are only to be used by the entities who have been allocated or paid for the right to use them. These organizations are referred to as the Primary Users (PUs). The restriction that these bands may only be utilized by their 'owners' has led to a large portion of the spectrum being unused in time and space. This unused spectrum is known as 'white space'. Cognitive Radio (CR) is a technology designed that allows users who do not have a license, known as Secondary Users (SUs), to opportunistically take advantage of vacant spectrum in cases where they are not in use by the PUs.

CRs sense the local environment for the presence of PU transmissions before they decide to transmit. This can be achieved via the construction of a Radio Environment Map (REM). A REM is a representation of the radio emissions environment and includes information such as primary user transmissions, which

S. Mc Gonigle • Q. Wang • M. Wang • A. Taylor • E.O. Nuallain (✉)
School of Computer Science and Statistics, Trinity College Dublin, Dublin, Ireland
e-mail: eamonn.onuallain@scss.tcd.ie

can be gathered from various sources including from the CRs themselves, and local radio emissions regulations [1]. CRs can consult the local REM to avoid interference with primary user transmissions.

Although the local REM information gathered by the cognitive radio is helpful, interference still may occur due to the hidden node problem. The Hidden Node Problem arises when, due to incomplete spectrum occupancy information, SU transmissions interfere with PU receivers. However by making use of the REM; complete spectrum occupancy information is made available to all members of the Cognitive Radio Network (CRN). Thereby resolving the Hidden Node Problem.

The global REM is constructed by having each node in the CR network provide their local REM information to a *data collector* which combines their spectrum sense information to form the global REM [2]. The availability of global REM information allows the secondary nodes to prevent the hidden node problem by checking for primary user transmissions that are outside their local spectrum sensing range.

The process of combining the SU spectrum sense information after it has been collected is known as *data fusion* [3]. This project will improve the accuracy of the global REM during *data fusion* by filtering out data supplied by secondary users which has a low trust value.

Although CRNs that utilise global REMs are infrastructure based, the spectrum sensing task required for accurate REM is distributed among CRs. The dependence of the CR network on the co-operation of these individual nodes introduces vulnerability as nodes may supply false information in order to disrupt the system. For example, a malicious secondary user might disrupt the fair allocation of spectrum by providing false local spectrum sensing information to the REM server. This is known as a *False Feedback Attack* [2]. For example a secondary node might carry out this type of attack in order to retain its bit rate for a longer period of time than would otherwise be allowed under a fair allocation system. The primary concern here is that this would lead SUs to fail to sense the presence of PU transmissions, leading to interference.

20.2 Trust/Reputation Based Security for Cognitive Radio Networks

In Kephart's vision of autonomic computing, the self-protecting property requires that the system automatically defends itself against malicious attacks [4]. In an attempt to move one step closer to this vision; this research will take inspiration from a mechanism used by human social networks where those who behave maliciously are excluded from the community. We hope to use a similar method of dealing with liars in a CR network. This involves taking the social concepts of trust and reputation and applying them to a CR technology.

As an example, where a community of fishermen shares the limited natural resource of fish population using a quota system then a lone fisherman may find it desirable to exceed their assigned quota and take more than their fair share; thereby increasing their profit margin. However, if too many fishermen do this then the resource becomes depleted to the point of being useless. This scenario is an example of a *Tragedy of the Commons* [5].

False feedback attacks on CR networks could also lead to a *Tragedy of the Commons*. Wireless spectrum is a natural resource that can be exploited in a similar fashion to any other natural resource [6]. Human society prevents the *Tragedy of the Commons* by applying *reputational pressure* [5] in order to discourage 'cheating'. For instance, it could be said that a fisherman was an over-fisher so that purchasers would not purchase his fish. The distinction between false feedback attacks and other *Tragedy of the Commons* examples is that the community members are non-human. This research attempts to apply the human idea of reputation to CR networks. This is achieved by assigning each node a trust value that can be adjusted appropriately whenever the node conforms to regulations or misbehaves. Whenever a CR supplies false information too often it is added to a black list and excluded from the community; much like the rogue fisherman.

Computational Trust is the application of the human concept of trust to networked computer systems in order to improve security. There is some difficulty in transferring this idea from social science to computer science because the social science notion of trust is subjective whereas computational trust must be based on well-defined metrics [7]. To illustrate, a node in a network will not be able to sense instinctively in human fashion if a neighbor is trustworthy, but it can query neighbors to determine if the node has been reliable in the past.

Measures of trust are either certificate-based or behavior based. Traditional certificate-based models use a Public Key Infrastructure along with a centralized Certificate Authority to verify if the data being transmitted is reliable. Behavior-based trust models evaluate trust by measuring of the level of positive co-operation between nodes in the network. This is carried out independently or sometimes as a community [8].

In Radio Environment Mapping for CR networks; we cannot rely on certificate-based trust alone. *Sensory manipulation attacks* confound the CR by manipulating the radio frequency that the radio device sees [9]. Attacks such as *sensory manipulation attacks*, cannot be secured cryptographically because these attacks involve manipulating the radio frequency rather than authenticated data. This kind of manipulation can only be detected by finding a way to debunk beliefs that do not make sense [9]. This research attempts to achieve this through a kind of '*common sense*' by testing beliefs regarding the radio environment of a particular node against those of nodes that are known to be trustworthy. Another reason for the introduction of non-traditional security methods is that certificate authorities (CAs) do not necessarily monitor the behavior of the node after it has been granted the certificate. A node may behave long enough to register with the CA and then misbehave later. Also, a large scale CR network will likely have a high volume of new radios registering with the network so there would be no chance to thoroughly screen each user before granting them a certificate.

A method proposed by [8] implements trust based security for what we refer to as a 'community-based network'. This method nominates the most trusted in the community to act as a 'base station' node. This node is then responsible for the distribution of public and private keys among the nearby nodes in the community [8].

This method alone does improve on conventional security methods but there is still room for improvement. We do this by employing the concept that is central to trust, namely, confidence. This differentiates between trust gained over a few exchanges and trust that is gained over a much longer period of time [7]. In an ad-hoc network nodes will be entering and leaving the network regularly. This hampers the establishment of time-tested trust.

Trust management systems maintain a *trust record*. The record is built from trust information gathered from direct observations and recommendations [7]. In a community-based approach this data would be held at the node which had the highest trust in the community. This approach may however be ineffective because this node itself may not have a very high trust value coming from, for example, a sparsely populated CR network.

20.3 Centralized Trust Management

Cognitive radio networks can either be centralized or decentralized. The centralized approach is achieved by connecting the secondary users to secondary base stations which are in turn connected via a wired backbone [10]. This section will discuss the attributes of the centralized and decentralized approaches and based on this we decide on an approach for our experiment.

The trust management systems described in [7] and [8] are both designed for ad-hoc community type networks. As mentioned before the disadvantage of this approach is that in an ad-hoc network, nodes frequently enter and leave. This behavior is detrimental to the development of a trust-based security system as there is less opportunity to establish the long term relationships between nodes that produce accurate trust values.

A centralized approach to trust management would involve holding trust values for each node in one location. This would differ from typical trust systems as trust recommendations would be requested from a central server rather from neighboring nodes. This is a characteristic of a reputation-based system as opposed to a trust-based system [11].

An ad-hoc cognitive radio network structure allows for 'On-Path' attackers. On-Path attackers are more dangerous than 'Off-Path' attackers because they have the ability to remove and alter traffic being transmitted whereas Off-Path attackers are only able to spoof data while being unable to observe traffic [9]. If a central server is responsible for maintaining the radio environment map then the data can be collected from the participating radios without any need to give them permission to read others. This makes a centralized approach to cognitive radio networking more resistant to On-Path attacks than an ad-hoc structure.

As a result it was decided to implement a centralized rather than an ad-hoc trust management approach because of the greater potential to build long-term trust and the ability to hold trust data in a fixed location.

20.4 Trust Model

This project evaluates the trustworthiness of a node by testing if the information supplied by that node for coherence with the information supplied by neighbors who are known to be trustworthy. Whenever a node supplies spectrum sense information to the central REM server; the software checks if the data corresponds with that data supplied by nodes that are in a nearby location. The server takes the intersecting area within that is within range of both nodes and checks if it is the same. If two or more nodes report conflicting information, then it can be deduced that at least one of them could be attempting a *false feedback attack*.

If a node with established trust disagrees with a newer member, then the newcomer will have its trust value decremented. If the trust value of a node goes below a certain threshold, then that node is added to a blacklist. Any node that is on the blacklist is excluded from the network. If the trust value of the node goes above a certain threshold then that node is added to a white list. Whenever a node is on the white list then the spectrum sense information supplied by that node is used during the *data fusion* stage in the construction of the global REM.

The system of reputation-based collaborative spectrum sensing described in [12] differs from the model implemented in this research as it adjusts node reputation based on its correspondence with the final global REM decision where as our model adjusts node reputation based on its correspondence with a more trusted neighbor's view. The weakness of the model described by this paper is that it requires trustworthy nodes to be present in the network at the beginning in order for trust values to propagate throughout the network. However, trust values can be adjusted more quickly if there is no need to wait until the global REM is initialized.

Another difference between the system proposed in this paper and conventional trust systems is that derived trust always involves communication with the server rather than just another SU. Nodes request trust recommendations from other nodes as shown in Fig. 20.1. In the model described in this research, the REM server will determine if a node is trustworthy by checking if the data provided by an unfamiliar node corresponds with the data provided by a more trusted node. The nodes can also request recommendations from the central server if they need trust information for non-Radio Environment Mapping purposes.

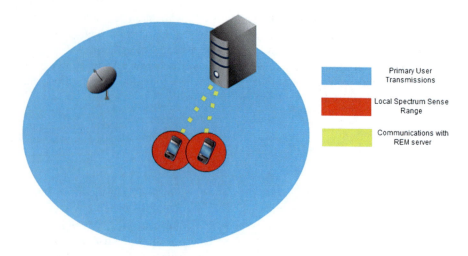

Fig. 20.1 Two nodes with overlapping coverage regions report to the server

20.5 Implementation

The simulation for this project was built in Java. The software developed for this experiment enables the number of desired nodes needs to be input and then generates the location of the nodes randomly. It is the responsibility of the individual nodes to report necessary information including its identity number and its location relative to a server during every process. Every node has a unique identity number and a trust value that ranges from 0 to infinity. The initial trust value of every node is 20, which is increased or decreased based on its behavior. The node will be added in the white list when its trust value is more than 30 and it will become a member of blacklist when the trust value falls below 5. Once a node has entered the white list or black list, it remains there permanently.

Each node is defined as being either malicious or co-operative at the outset. We specify that co-operative nodes always report the correct information to the server whereas malicious nodes always supply false information.

Each new node is assigned a new identity number when it enters the network. Position information from a new node will be compared with the information provided by its surrounding trustworthy nodes in order to determine if the new node is co-operative. However, if the neighbour's view of a new node is neither in white list nor black list, all information should be aggregated before making a decision. The new node will be marked as trustworthy when the information from it is consistent with more than half of other nodes; otherwise the new node is labeled a 'liar'. In some cases there will be no nearby nodes when the new node enters the network. In this case its trust value will maintain its initial value. The server receives reports and requests from nodes and then checks if the node can be trusted. The white list and black list mentioned above are held in the memory of the server which has the ability to recognize nodes according to nodes' report as well as their behaviour. The server constructs a REM from nodes in the white list.

In order to simplify the experiment the radio environment was divided into a rectangular section occupied by primary transmissions and a neighbouring rectangular section that was unoccupied. This is a gross simplification of a real life REM but it is enough to demonstrate the effectiveness of our trust system. In addition, a node can join or leave the test environment at any time. The test procedure is as follows:

1. Input the number of co-operative and malicious nodes into the user interface. Then the initial number of nodes is shown on the interface.
2. All nodes will report to server when nodes are transmitting. And the result of every round of processing will be shown on the user interface.
3. The trust value of a node will increase or decrease by 1 after every report and the server records all values.
4. The test is complete when the processing outcome does not change.

It is evident the lower the overall percentage of malicious nodes in the network, the greater the percentage of these malicious nodes that will be detected by the system. The average trust level across the entire network will decrease when more malicious nodes enter the network.

As discussed earlier, the server has the ability to construct the global REM based on the information provided by the nodes in the white list. Figure 20.2 shows a series of environment including the designed one and real environment detected by server. We design the environment is 100 × 100 spatial units, which is divided into two equal areas: PU network and vacant network space. However, the server does not have information about the primary network. The nodes in the white list send position information and their own trust value to server.

The first diagram below shows the hard coded 'real' spectrum environment that the server is supposed to measure. The three diagrams below it show the REM picture that was constructed by the server when the simulation was run with varying ratios of malicious nodes in the network (Fig. 20.3 and Table 20.1).

From the result shown, the REM server can accurately identify the REM when a high percentage of the malicious nodes have been detected. However, when an increasing number of malicious nodes manage to get on to the white list, it is clear that the accuracy of the server's REM construction suffers.

20.6 Discussion

Our method enables the REM server in a CR network to evaluate nodes' trustworthiness before allowing them to transmit. This clearly improves the security of the network. As we know, it is possible that nodes may misbehave after it obtains a certificate of trustworthiness. Our current approach cannot empower the server to keep checking the white list in order to find potentially malicious nodes. Our future

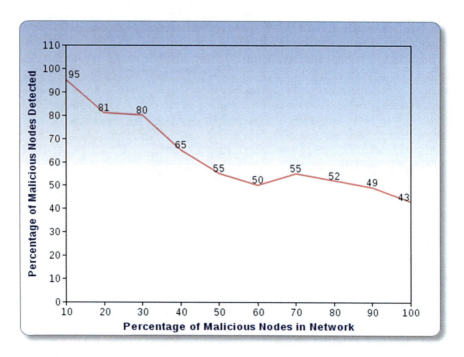

Fig. 20.2 Plot of percentage of malicious nodes detected versus percentage of malicious nodes in the network

work would implement this function by applying a more advanced algorithm that re-evaluates the trust of nodes that are already on the white list.

We have not yet solved the problem of how to evaluate the information provided by nodes that are very close to the border between PU coverage area and the vacant spectrum. The server may allow malicious nodes to enter the primary network without enough information on them in order to establish their reputation. As a result security may be negatively impacted because of these nodes joining the network.

It should be noted that more accurate results will be achieved if more co-operative nodes than malicious ones are input at the initial stage. Otherwise the result may be less reliable or even wrong when the number of malicious nodes is equal to or more than co-operative ones.

Currently the simulation software outputs the results in text format. An improvement to the test software would be to provide the user with a graphical representation of the statistics generated. For example the software could display the positions of the nodes during the simulation and generate graphics of the global REM afterward.

20 Reputation-Based Trust Management for Distributed Spectrum Sensing

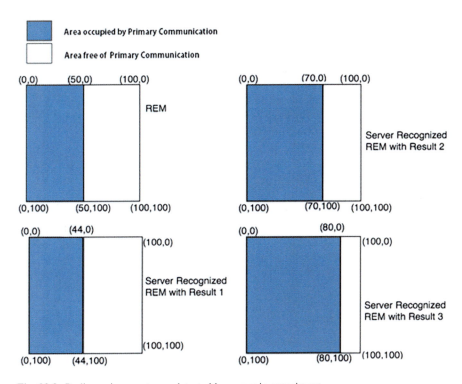

Fig. 20.3 Radio environment map detected by server in experiment

Table 20.1 Experimental results

	0 % of nodes malicious	3 % of nodes malicious	33 % nodes malicious
Co-operative nodes	300	300	300
Malicious nodes	0	10	30
Percentage malicious nodes detected	N/A	95 %	80 %

20.7 Conclusions

By distributing the task of spectrum sensing across the community of nodes instead of performing it locally, the CR network can address the hidden node problem more comprehensively. However, as more responsibility is distributed across the CR community there is more potential for members to abuse that responsibility by providing falsified spectrum sense information. This problem is not exclusive to CR but is inherent to distributed systems in general. The benefits of pooling the resources of all the nodes in the network are outweighed by the risk that malicious nodes could disrupt the system by lying to their neighbours.

Trust-based security methods can reduce these kinds of threats by providing the community with a mechanism to collaboratively identify and exclude misbehaving nodes.

As computer systems become more decentralized there will be a greater need for these kinds of security measures. It is likely that trust and reputation based security will need to become more advanced in order to adapt to the more sophisticated forms attack that come with more sophisticated kinds of node interdependence within the CR network.

This research has shown by means of a simple experiment that a reputation system can effectively reduce the number of malicious nodes in a network. However, this is by no means a complete solution. In a similar fashion to the concept of reputation in human social networks, the system becomes less effective whenever there is a particularly high ratio of malicious members.

In order for the trust system examined here to be effective, there must be a core group of highly trusted nodes present in the network so that derived trust can propagate to other nodes.

References

1. E.O. Nuallain, A proposed propagation-based methodology with which to address the hidden node problem and security/reliability issues in cognitive radio, in *Wireless Communications, Networking and Mobile Computing, 2008. WiCOM'08. 4th International Conference on Wireless Communications, Networking and Mobile Computing* (IEEE, 2008), Dalian, China, pp. 1–5
2. Y. Zhao, D. Raymond, C. da Silva, J.H. Reed, S.F. Midkiff, Performance evaluation of radio environment map-enabled cognitive spectrum-sharing networks. in *Military Communications Conference, 2007. MILCOM 2007. IEEE* (IEEE, 2007), Orlando, Florida, pp. 1–7
3. R. Chen, J.M. Park, Y.T. Hou, J.H. Reed, Toward secure distributed spectrum sensing in cognitive radio networks. Commun. Mag. IEEE **46**(4), 50–55 (2008)
4. J.O. Kephart, D.M. Chess, The vision of autonomic computing. Computer **36**(1), 41–50 (2003)
5. B. Schneier, *Liars and Outliers: Enabling the Trust that Society Needs to Thrive* (Wiley, Indianapolis, 2012)
6. New Scientist, Wireless spectrum: a hidden natural resource (2009), http://www.newscientist.com/blogs/shortsharpscience/2009/05/wireless-spectrum—a-precious-1.html
7. K.R. Liu, B. Wang, *Cognitive Radio Networking and Security: A Game-Theoretic View* (Cambridge University Press, Cambridge, MA, 2010)
8. S. Parvin, F.K. Hussain, Trust-based security for community-based cognitive radio networks. in *Advanced Information Networking and Applications (AINA), 2012 I.E. 26th International Conference on* (IEEE, 2012), Fukuoka, Japan, pp. 518–525
9. T.C. Clancy, N. Goergen, Security in cognitive radio networks: threats and mitigation. in *Cognitive Radio Oriented Wireless Networks and Communications, 2008. CrownCom 2008. 3rd International Conference on* (IEEE, 2008), Singapore, pp. 1–8
10. Q.H. Mahmoud, *Cognitive Networks* (Wiley, 2007), Chichester, West Sussex, pp. 57–71
11. A. Jøsang, R. Ismail, C. Boyd, A survey of trust and reputation systems for online service provision. Decis. Support Syst. **43**(2), 618–644 (2007)
12. K. Zeng, P. Paweczak, D. Cabric, Reputation-based cooperative spectrum sensing with trusted node's assistance. Commun. Lett. IEEE **14**(3), 226–228 (2010)

Chapter 21
Ontology Based Multi-Agent Intrusion Detection System for Web Service Attacks Using Self Learning

Krupa Brahmkstri, Devasia Thomas, S.T. Sawant, Avdhoot Jadhav, and D.D. Kshirsagar

Abstract Web Services (WS) have become a significant part of the Internet. They employ many features, each of them having specific drawbacks and security threats that are being exploited currently. According to current market researches majority of cyber attacks/exploits are done on these vulnerabilities in WS. Some are direct head on attacks while others are highly coordinated ones. To detect these attacks so that their further attempts can be prevented, highly intelligent Intrusion Detection Systems (IDS) are required. This can be done by having vast databases with high update frequencies or by employing a self learning ontology. Since, rules cannot be added to the database every minute and hence the ontology is preferred since attacks are of varying nature and new forms of attacks arise every day. For coordinated attacks, a single, stand alone IDS's becomes obsolete here. Hence the use of Distributed Intrusion Detection Systems (DIDS) along with firewalls is essential. The communication between these IDS's can be done using agents or any set standard of communication between these IDS's. On recognition of an attack on a single member or number of members of the DIDS System rules are added to the ontology knowledge base and learning occurs. This is the basic idea of an ontology based DIDS. The objective is to detect multiple kinds of attacks with good efficiency in least possible time practically.

K. Brahmkstri (✉) • D. Thomas • S.T. Sawant • A. Jadhav • D.D. Kshirsagar
Department of Computer Engineering and Information Technology, College of Engineering Pune, Pune, India
e-mail: brahmkstrikp09.comp@coep.ac.in; thomasda09.comp@coep.ac.in; sts.comp@coep.ac.in; jadhavavdhoot7@gmail.com; ddk.comp@coep.ac.in

21.1 Introduction

Most of the interactions happening today over the Internet are through Web Services. Along with the advantages offered by them, many drawbacks also exist. Approximately 75 % of cyber attacks/exploits are made on these vulnerabilities in WS. This is because the semantic data of WS are publicly available on the Internet.

Various attacks include the Mitnick attack, the semantic DoS (Denial of Service) attack, Application attacks, CDATA and XML attacks and SOAP attacks. IDS's can be used in these circumstances. Intrusion Detection System (IDS) is an application that monitors activities of the network and checks for the various attacks or potential harmful states of a host in the network and notifies the concerned system.

Traditional IDS's used taxonomies, rule based systems that required to be updated periodically. Looking at the current trend of attacks being generated daily, the frequency of updation is required to be at least once every hour. This is not feasible since users do not update with such a frequency and the party responsible for releasing updates, do not release that frequently. Hence we use a self learning ontology. Also, taxonomies are applicable only to a particular environment. Any modification to a situation will lead to a new taxonomy. Ontology helps overcome this by making taxonomies generic.

Agents are intelligent software objects used for decision making. They are capable of autonomous actions and are sociable, reactive and proactive in an information environment. Agents can answer queries, retrieve information from servers, make decisions and communicate with systems, other agents or with users. Use of agent-based systems enables us to model design and build complex information systems. An Intrusion detection system involving multi-agents makes the system strong, flexible and prevents it from single point of failure. It distributes the resources and task and hence each agent has its own independent functionality, so it makes the system perform work faster. Multi-Agents make the resilience of the system strong and thus ensure its safety.

21.2 Literature Survey

Vorobiev and Han [2] illustrates on some attacks possible on WS's. It is very detailed on the attacks mentioned and what other threats semantic data bring about. The methods of attack described here help to know about the nature of attacks further helping to design a method for detection. But this paper does not deal with an actual implementation though a proposed model is mention (a prevention system combined with a detection system).

Ye, Bai and Zhang [4] illustrates an agent based distributed IDS. Their method uses six agents particularly Monitor Agent, Analysis Agent, Executive Agent, Manager Agent, Retrieval Agent, and Result Agent. The former four agents are static agents that are inquiline on hosts, while the latter two are mobile agents that can travel among hosts. The monitor agent being the one given with the main task of monitoring process flows and network states of the peers in the P2P system. The advantage in using P2P is that the IDS is guaranteed to be distributed hence dealing with a broader class of detection. This paper helps to make the IDS having highly independent modules, which is required in reducing false alarms and their detection. But then again there is no particular implementation or proposed model even though illustrated model is highly educative.

Abdoli and Kahani [5] deals with a simulation of their proposed model of a distributed IDS using an agent based system. This model has two agents, an IDS agent and a master agent. The IDS agent this report behaves exactly like an IDS, detects a suspicious status and reports to the master agent. This master agent evaluates on the distributed level and proceeds further. Protégé was used to simulate the ontology and JENA for communication between agents. This paper too deals with a simulation and has no actual implementation. Also the IDS agent could have been split into more agents, since it the IDS agent is the IDS itself (although only for the local computer) giving the need that agents were introduced into this model as a mandatory requirement. The performance factors were measured on the KDD cup 99 dataset and the factors measured were the CPE, False alarms, and detection of DoS attacks.

Wei, Xu, Chen and Chaochun Xu [3] deals with similar aspects as the ones mentioned above. The topic that makes this paper stand out is that it discusses about privileged programs that none of the above have talked about. Since network attacks are phased and gradually the attackers increase their permission level. This is done by gaining access to privileged programs like remote computer accesses etc. This paper proposes monitoring of these programs to know if they are compromised or not. But then again there is no implementation cum proposed model, only a case study.

In the early stages of using agents in IDS a study named Java Agents for meta-learning [11] was conducted in which it combines intelligent agents and data mining techniques. When applied to the ID problem, an association-rules algorithm determines the relationships between the different fields in audit trails, while a meta-learning classifier learns the signatures of attacks. Features of these two data mining techniques are extracted and used to compute models of intrusion behavior.

By late 90s DARPA had defined the Common Intrusion Detection Framework (CIDF) as a general framework for IDS development. The Open Infrastructure [12] comprises a general infrastructure for agent based ID that is CIDF compliant. This infrastructure defines a layered agent hierarchy, consisting of the following agent types: Decision- Response agents (responsible for responding to intrusions), Reconnaissance agents (gather information), Analysis agents (analyze the gathered information), Directory/Key Management agents, and Storage agents. The two later provide support functions to the other agents.

In [13], a general Multi-Agent System (MAS) framework for ID is also proposed. Authors suggest the development of four main modules, namely the sniffing module (to be implemented as a simple reflex agent), the analysis module (to be implemented as several agents that keeps track of the environment to look at past packets), the decision module (to be implemented as goal-based agents to make the appropriate decisions), and the reporting module (to be implemented as two simple reflex agents: logging and alert generator agents). These components are developed as agents:

- The sniffing agent sends the previously stored data to the analysis agents when the latter request new data. One analyzer agent is created for each one of the attacks to be identified. They analyze the traffic reported from the sniffing module, searching for signatures of attacks and consequently building a list of suspicious packets.
- Decision agents are attack dependant. They calculate the severity of the attack they are in charge from the list of suspicious packets built by analyzer agents. Decision agents also take the necessary action according to the level of severity.
- Finally, the logging agent keeps track of the logging file, accounting for the list of suspect packets generated from the decision agents. On the other hand, the alert generator agent sends alerts to the system administrator according to the list of decisions.

Synergistic and Perceptual Intrusion Detection with Reinforcement in a Multi-Agent Neural Network (SPIDeR-MAN) is proposed in [14]. Each agent uses a SOM and ordinary rule-based classifiers to detect intrusive activities. A blackboard mechanism is used for the aggregation of results generated from such agents (i.e. a group decision). Reinforcement learning is carried out with the reinforcement signal that is generated within the blackboard and distributed over all agents who are involved in the group decision making.

A multi-agent IDS framework for decentralized intrusion prevention and detection is proposed in [15]. The MAS structure is tree-hierarchical and consists of the following agents:

- **Monitor agents:** capture traffic, preprocess it (reducing irrelevant and noisy data), and extract the latent independent features by applying feature selection methods.
- **Decision agents:** perform unsupervised anomaly learning and classification. To do so, an ant colony clustering model is deployed in these agents. When attacks are detected, they send simple notification messages to corresponding action and coordination agents.
- **Action agents:** perform passive or reactive responses to different attacks.
- **Coordination agents:** aggregate and analyze high-level detection results to enhance the predictability and efficiency.
- **User Interface agents:** interact with the users and interpret the intrusion information and alarms.
- **Registration agents:** allocate and look up all the other agents.

A MAS comprising intelligent agents is proposed in [6] for detecting probes. These intelligent agents were encapsulated with different AI paradigms: support vector machines, multi-variant adaptive regression, and linear genetic programming. Thanks to this agent-based approach, specific agents can be designed and implemented in a distributed fashion taking into account prior knowledge of the device and user profiles of the network.

By adding new agents, this system can be easily adapted to an increased problem size. Due to the interaction of different agents, failure of one agent may not degrade the overall detection performance of the network.

21.3 Proposed Model

The proposed architecture is similar to many models mentioned before, but is more optimised in terms of parallelization and inter agent communication.

The reason that this paper's architecture is optimized is that we have used an adequate number of agents for optimal parallelization as well as since the number of agents are relatively small and the resources required by them aren't common, the deadlock situation does not occur. The inter-agent communication flow makes our architecture look serialized but is actually a pipeline.

According to Gruber [1], ontology defines a set of representational primitives with which to model a domain of knowledge or discourse. The representational primitives are typically classes (or sets), attributes (or properties), and relationships (or relations among class members). Therefore, ontology is designed for the purpose of enabling knowledge sharing and reuse between entities within a domain. In this research, these entities are various agents in our IDS.

21.3.1 Monitor Agent

Monitor Agent is responsible for monitoring activities and host statuses by looking at appropriate data structures. It captures network packets and forwards it to the pipelined analysis agent for further processing.

21.3.2 Analysis Agent

This agent has multiple instances running at the same time. Each connection is uniquely identified by the remote host's network properties. The information sent by the Monitor Agent here is either sniffed or parsed depending on the nature of the packet. The processing is done with the help of a knowledge base. On successful processing of this information, it is then decided if it is malicious or not. If it is malicious, then the Executive Agent is called by sending appropriate parameters. Before doing the analysis, Analysis Agent checks with the Manager Agent if there was any malicious activity from the host in the recent past. If so, then the connection is directly terminated and there is no need for parsing.

21.3.3 Executive Agent

The final task delivery is done by this system. The Executive Agent is in charge of warning other nodes in this DIDS. It also terminates connections with malicious intents and attempts recovery. Also when the Analysis Agent detects malicious packets but the pattern of intrusion is new then the Executive Agent adds this new pattern to the Knowledge Base (Artificial Intelligence).

21.3.4 Manager Agent

The warnings sent by the Executive Agents of other nodes are received by this Manager Agent. The Manager Agent has a temporary database, in which these warnings are stored, that exists for the programs lifetime (essentially this means that the database is made every time the system is turned on). The function of this Manager Agent is to supply information of host performing malicious activities to the Analysis Agent.

21.3.5 Knowledge Base

It consists of attack patterns, malicious code patterns that are used by the analysis agents and updated by the executive agent. That means it contains certain rules which help the Executive Agent decide upon as to which data is malicious and thus generate warning.

21.4 System Implementation

In this paper, detection methods for the Smurf and Mailbomb attacks are discussed in detail.

The Knowledge base consists of an Attack Ontology and instances of it which represent many attack types. The ontology has data properties which represents the attributes of a connection. Instances contain values for these data properties, pointing to a particular attack. Hence the same ontology is reused between entities of the same domain. The same ontology is considered across different IDS's for the attack detection. The ontology and instances are created by using a tool called Protégé.

The Monitor Agent aggregates the packets received in a connection between two nodes. Then, it extracts attributes of the connection. These attributes are then passed on to the Analysis Agent. For research purpose, KDD Cup 99 has been taken for traffic analysis at the Monitor Agent (Fig. 21.1).

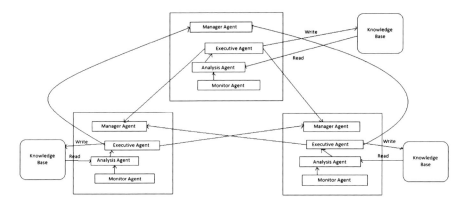

Fig. 21.1 Proposed architecture

The Analysis Agent now compares the attributes it has received against the instances stored in knowledge base. This is done by parsing the ontology by using the Jena API used to interface java with ontologies. Once the ontology is parsed (takes place only once), the instances are queried using the attributes received to get a quick result. If there is no return result, it implies that this connection attribute pattern isn't recognizable by the system and hence this pattern is treated as an attack temporarily and fed into a new attack database, and further occurrences of this pattern is blocked (self learning) and fed into a reference database. The administrator or an expert person studies these two databases to classify it as normal or as an attack and hence creating an instance of it. It is referred as classifying since minor variations of attributes will not hinder detection. If there is a result returned, and, if it is a normal connection, then no further processing takes place. But if it is not a normal then the connection is forwarded to the Executive Agent which contains a rule set of how to deal with each attack contained in the database. Since Smurf and Mailbomb are DoS attacks they can be recognized by the size data exchanged between the source (attacker) and destination (node having IDS). Smurf has the source sending 1032/520/508 bytes of data per connection and the destination sends nothing back. And it is done so over the ICMP protocol using service ecr_i. Whereas, Mailbomb has source sending 2,599 bytes of data and destination sending 293 bytes of data. This is run over the TCP protocol using the SMTP service.

The Executive Agent, as mentioned before, has a rule set for reference on how to the system is how to react to a particular attack. In general it broadcasts message over a particular port number informing other nodes that it is under a particular attack along with its connection details and also decides to block the connection on the node under attack or not. In regard to the above mentioned attacks the executive agent broadcasts the name of the attack. Since it is a DoS attack other nodes shall drop all active connections with the affected node till it broadcast that the DoS has been lifted.

The Manager Agent stores all the messages it receives from Executive Agents of other nodes in a temporary database. The Analysis Agent refers to this database before it starts parsing/querying. In case there is an entry of the current connection then it directly forwards to the executive agent the attack type. The Manager Agent modifies its database on basis of messages received from executive agents. In case of a failure and this database is lost, it just copies it from the nearest node.

The messages exchanged are MD5 hashed.

21.5 Results

The factors that are taken into explaining the system's results are true positives, true negatives, false positives, false negative and its measures. These parameters help us to measure the following:

$$\text{Precision (P)} = tp/(tp+fp)$$
$$\text{Recall (R)} = tp/(tp+fn)$$
$$\text{Accuracy (A)} = (tp+tn)/(tp+fp+tn+fn)$$
$$\text{Fmeasure (F)} = 2^*P^*R/(P+R)$$

Our sample database had data of 294,932 connections which had packets that were normal and also of malicious intent. From this collection we tested for the following two attacks to see how well the system could classify attacks and also tested our system against normal packets to see how well the system could differentiate between normal and attack scenarios.

For Smurf, our system performed in the following manner:

tp = 163,475 (correctly identified), fp = 3 (falsely identified),
fn = 616 (incorrectly allowed), tn = 130,838 (correctly identified as not Smurf)

Hence:

P = 99.99 %, R = 99.62 %, A = 99.79 %, F = 99.80 %

For Mailbomb, our system performed in the following manner:

tp = 5,000 (correctly identified), fp = 0 (falsely identified),
fn = 0 (incorrectly allowed), tn = 289,932 (correctly identified as not Mailbomb)

Hence

P = R = A = F = 100 %

Now in the final test (as mentioned before, it was differentiate between attack and normal packets), we observed the following.

tp = 44,496 (correctly identified), fp = 762 (falsely identified),
fn = 0 (incorrectly allowed), tn = 249,674 (correctly identified as attacks)

Hence

P = 98.31 %, R = 100 %, A = 99.74 %, F = 99.15 %

21.6 Conclusion and Future Work

In this paper, the proposed model of a Network based IDS and also implementation is explained. Also, various IDS's are studied. It shows that there is considerable variation in the factorization each model has used and each model can be used for specific type of attack as well as for a specific system. The paper introduces an ontology based Multi- Agent IDS for web service attacks. All the agents are assigned different functions so that proper decision regarding the attack can be taken. The model applies self learning and distributes the data among all the hosts using minimum number of agents required. Thus, this detection system minimizes the overload of the use of its multiple components. It is found that the detection accuracy for Smurf attack is 99.79 % and Mailbomb attack is 100 % thus showing that Mailbomb attack is better detected, but, at the same time the system does not falsely identify an attack scenario as a normal scenario (fn = 0, last test), hence showing the effectiveness of the system.

The working explained here is based on an implementation scenario of an offline IDS. Future work would concentrate on implementing this system in an Online IDS scenario that will take real time traffic and analyze it. After this is done the aim is to make it an Intrusion Prevention System.

References

1. T.F. Gruber, A translation approach to portable ontologies. Knowl. Acquis. **5**(2), 199–220 (1993)
2. A. Vorobiev, J. Han, Security attack ontology for web services, in *Proceedings of the Second International Conference on Semantics, Knowledge, and Grid (SKG'06)*, 2006, Guilin, China, Paper 42, (6 pp.)
3. Mingjun Wei, Guangli Xu, Xuebin Chen, Chaochun Xu, Study on ontology-based intrusion detection, in *International Conference on Computer Application and System Modeling (ICCASM)*, 2010, Taiyuan, China, pp. V10-357–V10-359
4. D. Ye, Q. Bai, M. Zhang, Ontology-based knowledge representation for a P2P multi-agent distributed intrusion detection system, in *IFIPA International Conference on Network and Parallel Computing*, ed. by J. Cao, M. Li (IEEE Computing Society, Los Alamitos), pp. 111–118
5. F. Abdoli, M. Kahani, Ontology-based distributed intrusion detection system, in *Proceedings of the 14th International CSI Computer Conference, 2009 (CSICC 2009)*, Tehran, Iran, pp. 65–70
6. Y. Lasheng, M. Chantal, Agent based distributed intrusion detection system (ABDIDS), in *Second Symposium International Computer Science and Computational Technology (ISCSCT'09)*, 2009, Huangshan, P.R. China, pp.134–138
7. A. Razzaq, A. Hur, M. Masood, K. Latif, H. Farooq Ahmad, H. Takahashi, Foundation of semantic rule engine to protect web application attacks, in *Proceedings of International Symposium on Autonomous Decentralized Systems, (ISADS)*, 2011, Tokyo & Hiroshima, Japan, pp. 95–102
8. L. Frye, L. Cheng, J. Heflin, An ontology-based system to identify complex network attacks, in *IEEE International Conference on Communications (ICC)*, 2012, Ottawa, pp. 6683–6688

9. R.R. de Azevedo, E.R.G. Dantas, F. Freitas, C. Rodrigues, M.J.S.C. de Almeida, W.C. Veras, R. Santos, An autonomic ontology-based multiagent system for intrusion detection in computing environments. Int. J. Infonomics (IJI). **3**(1), (2010)
10. J. Undercoffer, A. Joshi, J. Pinkston, Modeling computer attacks: an ontology for intrusion detection, in *RAID,* LNCS, vol. 2820 (Springer, Berlin, 2003), pp. 113–135
11. S. Stolfo, A.L. Prodromidis, S. Tselepis, W. Lee, D.W. Fan, P.K. Chan, JAM: Java agents for meta-learning over distributed databases, in *Third International Conference on Knowledge Discovery and Data Mining*, 1997, Newport Beach, pp. 74–81
12. M. Reilly, M. Stillman, Open infrastructure for scalable intrusion detection, in *IEEE Information Technology Conference*, 1998, Syracuse, pp. 129–133
13. I.M. Hegazy, T. Al-Arif, Z.T. Fayed, H.M. Faheem, A multi-agent based system for intrusion detection. IEEE Potentials **22**(4), 28–31 (2003)
14. P. Miller, A. Inoue, Collaborative intrusion detection system, in *22nd International Conference of the North American Fuzzy Information Processing Society*, 2003, Chicago, pp. 519–524
15. C.-H. Tsang, S. Kwong, Multi-agent intrusion detection system in industrial network using ant colony clustering approach and unsupervised feature extraction, in *IEEE International Conference on Industrial Technology (ICIT 2005)*, 2005, Budapest, Hungary, pp. 51–56
16. S. Mukkamala, A.H. Sung, A. Abraham, Hybrid multi-agent framework for detection of stealthy probes. Appl. Soft Comput. **7**(3), 631–641 (2007)

Chapter 22
A Novel Key Exchange Protocol Provably Secure Against Man-in-the-Middle Attack

Abhijit Chowdhury, Shubhajit Nath, and Jaydeep Howlader

Abstract Diffie-Hellman (D-H) key exchange protocol, first published in 1976 is one of the most distinguished and widely used secured key exchange protocols. The protocol makes it possible to exchange a symmetric key between two parties over an unsecured communication channel without pre-agreement of any shared secret. The D-H key exchange protocol however requires a pre-set public parameters P (a large prime) and g a generator of $\mathbb{Z}_p$ to be public i.e. known to all users of the system. This paper proposes a key exchange mechanism without any public parameter. Unlike D-H key exchange mechanism our scheme uses encryption to transmit intermediate data during key exchange stage, thus it is non-vulnerable to man-in-the-middle attach.

22.1 Introduction

The last two decades have seen enormous growth in faster computer systems and digital communication systems with the developments of technologies of manufacturing cheap digital components required in manufacturing computer systems. Computing systems are widespread in areas starting from like Engineering, manufacturing, banking, military, education, hospital, aviation to household. Most systems are now a day's connected to local area network or wide area network. The volume of digital data has increased significantly, with such growth requirement of secured system to protect such information has grown phenomenally. Many systems like ATM machines to online transaction processing require real-time security

A. Chowdhury (✉) • S. Nath
Department of Computer Applications, NSHM College of Management and Technology, Durgapur, West Bengal, India
e-mail: chowdhury.abhijit@gmail.com

J. Howlader
Department of Information Technology, National Institute of Technology, Durgapur, West Bengal, India

to protect data as well as authenticate legitimate parties. This has evolved requirement of cryptographic protocols [1] to protect valuable and sensitive digital information. Cryptographic systems [2] are broadly classified as symmetric and asymmetric. In symmetric systems, the sender at encryption end and receiver at decryption end require identical key or a single key. While asymmetric cipher requires different keys for encryption and decryption. The major problem with such systems is mechanism or mode of exchanging key or keys between users without disclosing it to adversary. In cryptography, key exchange can be defined as a scheme that allows sharing of cryptographic key [3] between two or more legitimate users without being disclosing the keys to an adversary. Diffie-Hellman (D-H) [4] key exchange protocol, first published in 1976 is one of the most distinguished and widely used secured key exchange protocols over an unsecured channel. Section 22.1.1 presents a brief review of the D-H Key Exchange Protocol, which is significant to the present context. D-H key exchange protocol requires some public parameters known to the two parties willing for key exchange to execute the key establishment. In Sect. 22.2 of this paper, we have presented a novel approach of key exchange where the two parties need not pre-agree upon g or P.

22.1.1 Review of D-H

Alice and Bob two legitimate users wish to share a secret key over an unsecured channel for use in a symmetric key cipher. They agree to use two integers P, a large prime and g (a primitive root of P) at start up P and g are declared public to all users. Alice picks a secret integer a, a $\{\in\}$ $\mathbb{Z}_p$, which is not revealed to anyone, computes $k_a = g^a$ mod P and sends k_a to Bob over unsecured channel. At the same time Bob chooses a secret integer b, b $\{\in\}$ $\mathbb{Z}_p$ and computes $k_b = g^b$ mod P and sends k_b to Alice. In the next step Alice computes key $= (g^b)^a$ mod P and Bob computes key $= (g^a)^b$ mod P. an adversary can see values g^a, g^b but it is difficult for him or her to compute $(g^a)^b$ mod P.

22.1.2 Our Contribution

The proposed key exchange protocol is based on the hardness of factorization problem of big numbers. The two parties willing for key exchange start without any pre-agreed secrets. The protocol executes in two parts i.e. the setup stage and the key exchange stage. In setup stage, each party sets an encryption key and decryption key. The second stage makes the key exchange.

In Sect. 22.2, we have discussed our proposed key exchange protocol in further details. Section 22.3 contains analysis of the proposed scheme. In Sect. 22.4, we draw a conclusion.

22.2 The Scheme

Let Alice and Bob are two parties who wish to exchange a symmetric key over an unsecured channel.

We divide the scheme in two stages, the setup stage and the key exchange stage.

In the setup stage, two parties generate a separate set of private and public key, used to encrypt and decrypt the intermediate results during the key exchange stage.

22.2.1 The Setup Stage

Now we describe the setup stage. Alice chooses two prime numbers P and Q while Bob takes two prime numbers R and S. next Alice computes the followings:

1. Set $M = P^A Q^B$ where A, B are integers >1
2. Set $K = \phi(P)$
3. Set $L = \phi(Q)$
4. Set $Z = LCM(K^A, L^B)$
5. Compute $E_1.D_1 \bmod Z = 1$

In step 2 Alice computed $K = \phi(P)$ where that the function $\phi(P)$ is a totient [5] function.

Now, Alice sends (E_1, M) to Bob via open unsecured channel. Alice keeps (D_1, PQ) as secret to herself. Alice will use the public key [6] pair (E_1, M) to encrypt any intermediate data that she will be sending to Bob during the key exchange stage.

Similarly, bob computes:

1. Set $N = R^A S^B$ where A, B are integers >1
2. Set $K = \phi(R)$
3. Set $L = \phi(S)$
4. Set $Z = LCM(K^A, L^B)$
5. Compute $E_2.D_2 \bmod Z = 1$

Now, Bob sends (E_2, N) to Alice via open unsecured channel. Bob keeps (D_2, RS) as secret to him.

Bob will use the non-secret [7] key pair (E_2, N) to encrypt any intermediate data that he will be sending to Alice during the key exchange stage.

Alice, using the one-way trap-door [8] pair (D1, PQ) will decrypt any encrypted intermediate data send by Bob. Bob will use the one-way trap-door pair (D_2, RS) to decrypt the intermediate encrypted data sent by Alice.

Alice and Bob now agree upon an integer T ($T < M$, $T < N$) to use as a modulus. Now both Alice and Bob are ready for key exchange.

22.2.2 The Key Exchange Stage

The Key Exchange stage is divided into six steps (Tables 22.1 and 22.2).

Step 1:
In first step of key exchange, Alice chooses two secret integers a, h such that GCD(a, T) = 1 then computes the value $\alpha_A = [a^{(h)} \mod T]$ and encrypts it using the key pair (E_2, N) to get $[\alpha_A]^{E_2} \mod N$ and sends it to Bob.

At the other end Bob computes $\alpha_B = [b^{(i)} \mod T]$ then encrypts it with the key pair (E_1, M) to get $[\alpha_B]^{E_1} \mod M$ and sends it to Alice, where b and i are some random integers. While choosing b bob should take care that such that GCD (b, T) = 1.

Step 2:
In the second step Alice computes $g1 = [[\alpha_B]^{E_1} \mod M]^{D1} \mod PQ$. Similarly, in second step, Bob computes $g2 = [[\alpha_A]^{E_2} \mod N]^{D2} \mod RS$.

Step 3:
In third step of key exchange, Alice computes $G = (g1 \times \alpha_A) \mod T$. Similarly Bob computes $G = (g2 \times \alpha_B) \mod T$.

Step 4:
In step four, Alice computes $\beta_A = [G^{(h)} \mod T]$ then encrypts it as $[\beta_A]^{E_2} \mod N$ and sends to Bob. Similarly, Bob computes $\beta_B = [G^{(i)} \mod T]$ then encrypts it as $[\beta_B]^{E_1} \mod M$ and sends to Alice.

Step 5:
In step five Alice computes $K1 = ([\beta_B]^{E_1} \mod M)^{D1} \mod PQ$. Similarly, in step five Bob computes $K2 = ([\beta_A]^{E_2} \mod N)^{D2} \mod RS$.

Step 6:
In sixth and final step Alice compute $KEY = K1^{(h)} \mod T$. Similarly, in final step Bob computes $KEY = K2^{(i)} \mod T$.

22.3 Analysis

The proposed key exchange uses a novel approach of encrypting the intermediate data generated during key exchange stages. Both the participants Alice and Bob decrypt the encrypted data to get back the intermediate data generated and send by the other party during key exchange.

There is no need of using some pre-agreed prime and a generator integer to start the process of key exchange. The adversary who watches the information transfer over unsecured channel gets to know about encrypted data from which constructing G, h or i will be difficult. Unless the adversary knows G, it is not possible for him or her to setup man-in-the-middle attack.

Table 22.1 Shows the computational steps of the proposed scheme

Computational steps followed by Alice and Bob	
Alice	Bob
1. Alice chooses two secret integers a, h such that GCD(a, T) = 1 then computes the value $\alpha_A = [a^{(h)} \mod T]$	1. Bob computes $\alpha_B = [b^{(i)} \mod T]$ then encrypts it with the key pair (E_1, M) to get $[\alpha_B]^{E_1} \mod M$ and sends it to Alice, where b and i are some random integers
Alice computes $[\alpha_A]^{E_2} \mod N$ and sent to Bob, where a and h are some random integers	Bob computes $[\alpha_B]^{E_1} \mod M$ and sent to Alice
2. Alice computes $g1 = [[b^{(i)} \mod T]^{E_1} \mod M]^{D_1} \mod PQ$	2. Bob computes $g2 = [[a^{(h)} \mod T]^{E_2} \mod N]^{D_2} \mod RS$
3. Alice computes $G = (g1 \times a^{(h)} \mod T) \mod T$	3. Bob computes $G = (g2 \times b^{(i)} \mod T) \mod T$
4. Alice computes $\beta_A = [G^{(h)} \mod T]$ then encrypts it as $[\beta_A]^{E_2} \mod N$ and send to Bob	4. Bob computes $\beta_B = [G^{(i)} \mod T]$ then encrypts it as $[\beta_B]^{E_1} \mod M$ and send to Alice
5. Alice computes $K1 = ([\beta_B]^{E_1} \mod M)^{D_1} \mod PQ$	5. Bob computes $K2 = ([\beta_A]^{E_2} \mod N)^{D_2} \mod RS$
6. Alice compute $KEY = K1^{(h)} \mod T$	6. Bob computes $KEY = K2^{(i)} \mod T$

In setup stage Alice and Bob publishes (E_1, M) and (E_2, N) respectively, but with knowledge of such pairs the adversary cannot compute (D_1, PQ) or (D_2, RS) as factorization of M ($M = P^A Q^B$) or N ($N = R^A S^B$) is hard.

In key exchange stage Alice computes $\alpha_A = [a^{(h)} \mod T]$ Bob computes $\alpha_B = [b^{(i)} \mod T]$ the integer T is determined on the flow of the protocol and does not impose any constrain except T < M and T < N.

The output KEY is within $\mathbb{Z}^*_T$, that is even M and N are bigger than T the key value is limited within the range of [1–T].

From α_A it is difficult to determine a or h. similarly from α_B it is difficult to determine b or i.

In every session of key exchange resetting of values of P, Q and R, S will ensure that key replay attack is not possible in future session of key exchange.

22.4 Conclusion

The proposed key exchange protocol can generate much larger set of keys compared to conventional D-H key exchange protocol. Time and number of operations needed to identify the key in the proposed key exchange protocol is much larger. The proposed key exchange scheme is provably secure against man-in-the-middle attack and key replay attack. The sharing of pre-agreed prime and its generator between the users who requires a secured key exchange is also not required in the proposed protocol.

Table 22.2 An example of the scheme is shown

Example of the scheme with numeric values	
Alice	Bob
Encryption and decryption key generation	
$P = 11, Q = 3, PQ = 33$	$R = 7, S = 5, RS = 35$
$A = 2, B = 3$	$A' = 2, B' = 3$
$M = P^A Q^B$	$N = R^{A'} S^{B'}$
$M = 11^2 3^3 = 1{,}089$	$N = 7^2 5^3 = 6{,}125$
$K = \Phi(P) = 10$	$K' = \Phi(R) = 6$
$L = \Phi(Q) = 2$	$L' = \Phi(S) = 4$
$Z = LCM(K^A, L^B)$	$Z' = LCM(K'^{A'}, L'^{B'})$
$Z = LCM(10^2, 2^3) = 200$	$Z' = LCM(6^2, 4^3) = 576$
$E_1.D_1 \bmod Z = 1$	$E_2.D_2 \bmod Z' = 1$
$E_1 = 3, D_1 = 67$	$E_2 = 5, D_2 = 461$
Key exchange phase	
$a = 22, h = 9, T = 26$	$b = 16, i = 17, T = 26$
$\alpha_A = [a^{(h)} \bmod T] = 14$	$\alpha_B = [b^{(i)} \bmod T] = 22$
$[\alpha_A]^{E_2} \bmod N$	$[\alpha_B]^{E_1} \bmod M$
$[14]^5 \bmod 6{,}125 = 4{,}949$	$[22]^3 \bmod 1{,}089 = 847$
$g1 = [\,[\alpha_B]^{E_1} \bmod M\,]^{D_1} \bmod PQ$	$g2 = [\,[\alpha_A]^{E_2} \bmod N\,]^{D_2} \bmod RS$
$g1 = [847]^{67} \bmod 33 = 22$	$g2 = [4{,}949]^{461} \bmod 35 = 14$
$G = (g1 \times \alpha_A) \bmod T$	$G = (g2 \times \alpha_B) \bmod T$
$G = (22 \times 14) \bmod 26 = 22$	$G = (14 \times 22) \bmod 26 = 22$
$\beta_A = [G^{(h)} \bmod T]$	$\beta_B = [G^{(i)} \bmod T]$
$\beta_A = [22^{(9)} \bmod 26] = 14$	$\beta_B = [22^{(17)} \bmod 26] = 16$
$[\beta_A]^{E_2} \bmod N$	$[\beta_B]^{E_1} \bmod M$
$[14]^5 \bmod 6{,}125 = 4{,}949$	$[16]^3 \bmod 1{,}089 = 829$
$K1 = ([\beta_B]^{E_1} \bmod M)^{D_1} \bmod PQ$	$K2 = ([\beta_A]^{E_2} \bmod N)^{D_2} \bmod RS$
$K1 = (829)^{67} \bmod 33 = 16$	$K2 = (4{,}949)^{461} \bmod 35 = 14$
$KEY = K1^{(h)} \bmod T$	$KEY = K2^{(i)} \bmod T$
$KEY = 16^{(9)} \bmod 26 = 14$	$KEY = 14^{(17)} \bmod 26 = 14$

References

1. B. Schneier, *Applied Cryptography, Second Edition: Protocols, Algorithms, and Source Code in C* (Wiley). John Wiley & Sons, (1995)
2. A.J. Menezes, P. van Oorschot, S. Vanstone, *Handbook of Applied Cryptography* (CRC Press, 1996)
3. R.L. Rivest, A. Shamir, L. Adleman, A method for obtaining digital signatures and public-key cryptosystems. Commun. ACM **21**(2), 6–7 (1978)
4. W. Diffie, M. Hellman, New directions in cryptography. IEEE Trans. Info. Theory **22**(6), 34 (1976)
5. C.S. Turner, Euler's totient function and public key cryptography. http://web.cs.du.edu/~ramki/courses/security/2011Winter/notes/RSAmath.pdf
6. M.O. Rabin, Digitalized signatures and public-key functions as intractable as factorization. MIT/LCS/TR-212, (1979)
7. C. Cocks, A note on non-secret encryption. CESG report, (1973)
8. K. Schmidt-Samoa, A new rabin-type trapdoor permutation equivalent to factoring and its applications, in *IACR Cryptology Archive* (2005)

Chapter 23
Neutralizing DoS Attacks on Linux Servers

G. Rama Koteswara Rao and A. Pathanjali Sastri

Abstract Worldwide IT industry is shifting towards Service Oriented Architecture at a fast pace. To meet this emerging scenario, most of the organizations are adopting business models such as cloud computing that are dependent on reliable server platforms. Linux servers are well ahead of other server platforms in terms of security. This brings network security to the forefront of major concerns to an organization. The most common form of attacks is a Denial of Service attack. This paper focuses on mechanisms to detect and immunize Linux servers from DoS.

23.1 Introduction

Denial of Service attack is an attack that damages a server's hardware and software resources that is initiated by a person or any other system. These resources can be operating system data structures [2]. It makes a server unreachable and prevents end users accessing services of the server, modify system configuration information and can even destroy physical network components. These attacks disable a network, cause loss of data and results in financial losses to an organization. The risk of Denial of Service attack is unavoidable. DoS attacks are always malicious and illegal. Well-known popular web sites are repeatedly strike down by malicious hacker. To defeat detection the attacker can easily manipulate their traffic and the problem of identifying attack will be very difficult [6]. As per the survey conducted by FBI, these attacks are dreadful attacks in terms of financial losses for the

---

G.R.K. Rao (✉)
Department of Information Technology, V.R. Siddhartha Engineering College,
Vijayawada, India
e-mail: koti_g@yahoo.com

A.P. Sastri
Department of Computer Applications, V.R. Siddhartha Engineering College,
Vijayawada, India
e-mail: akellapatanjali@yahoo.com

organizations after information thefts [12]. As DoS attacks have become more regular, the DoS problem has inspired an avalanche of research into solutions [21].

This paper focuses on preventing DoS attacks from harming Linux servers. Common forms of Denial of Service (DoS) attacks include TCP SYN flooding attacks, TCP Sequence Number Attack, TCP Hijacking, ICMP Smurf attacks and Packet Spoofing.

23.2 Related Work

Wentao Liu discussed on DoS attack principle and some DoS attack methods are deeply analyzed and presented DoS attack detection technologies that includes network traffic detection and packet content detection [23].

Xinyu Yang focused on the typical DoS/DDoS attacks under IPv6, which includes the DoS attacks pertinent to IPv6 Neighbor Discovery protocol and DDoS attacks based on the four representative attack modes, they are respectively TCP-Flood, UDP-Flood, ICMP-Flood and Smurf [24].

Pukkawanna, S proposed a lightweight method to identify DoS attacks by analyzing host behaviors. This method is based on the concept of BLINd Classification or BLINC: without access to packet payload, without knowledge of port numbers, and without additional information other than what current flow collectors provide [20].

M. Voznak performed comparison of the DoS attack's efficiently which were tested both without any protection and then with implemented Snort and SnortSam applications as proposed in the solution [17].

Bin Xiao proposed to provide a method that detects SYN flooding attacks in a timely fashion and that responds accurately and independently on the victim side [4].

Yao Chen, S. Das, Pulak Dhar, Abdulmotaleb El Saddik, and Amiya Nayak proposed a novel scheme for detecting and preventing the most harmful and difficult to detect DDoS Attacks – those that use IP address spoofing to disguise the attack flow [25].

C.L. Schuba, I.V. Krsul, M.G. Kuhn, E.H. Spafford, A. Sundaram, D. Zamboni contributed a detailed analysis of the SYN flooding attack and a discussion of existing and proposed countermeasures. Introduced a new approach that offers protection against SYN flooding for all hosts belonging to the same local area network, with their independent operating system or networking stack implementation. It is highly portable, configurable and extensible [8].

23.2.1 Solution to Identify and Block the DoS Attacks in Linux Kernel Version 2.6 and Above

Current work focuses on the packet filtering rules that are defined in the firewall/router to identify and block the attacks. These rules monitor the network traffic, its source, its destination and its protocol type. This work focuses on

- Identifying attacking network and blocking it.
- To develop security measures on the Server.
- Analyzing number of times each IP connected to the Server.
- Monitoring Load on the CPU
- Check if the Server is flooded with SYN requests.
- Check the Server flooded with ICMP echo requests.
- Check if any DoS attack is targeting the Server.

23.3 TCP SYN Flooding Attacks

One of the most severe forms of attack is TCP SYN Flood attack because legitimacy of a client cannot be established during a TCP SYN Flood attack. Once the target host's resources are tired, no more incoming TCP connections can be recognized, thus denying further legitimate access [7]. During SYN Flood attacks, the attacker sends SYN packets with non existing source IP addresses [14].

In SYN attack the client uses faked IP address to sends SYN messages to the Server. The server sends an ACK message that is for no reason returned. The server uses up its resources in the process while waiting for the ACK message from the client. The server becomes slow or insensitive to the other clients when the server is loaded with ACK messages. Flooding spoofed SYN requests can easily fail the victim server's backlog queue, causing all the arriving SYN requests to be dropped [10]. These attacks are unsafe, put away the server and make the websites and networks on the server unapproachable. In the critical real-time services the server may be slow or shutdown or kills valuable resources due to flooding of packets by SYN Flood attacks [15].

23.3.1 Detecting a SYN Flood Attack

When sending a vicious packet from compromised client, this client cannot receive a SYN/ACK packet from server [13]. The SYN Flood attack will not allow the Server to receive expected ACK code. The symptom of SYN Flood attack on the server is that the performance of the server will be slow. For Example the web site on the server will take long time to load or loads some elements of the page but not all. The server could be under SYN Flood attack, if the attack is directed with large number of SYN_RECV packets from a single IP address. This problem can be solved by adding the IP in firewall to stop the attack.

23.3.2 Defending from a SYN Flood Attack

SYN flooding targets end hosts rather than attempting to weaken the network ability, it seems logical that all end hosts should implement defences, and that network-based techniques are an optional second line of defense that a site can employ [22]. The following steps are employed to defend from a SYN Flood Attack.

- Allow the server to avoid reducing connections when the SYN queue fills up.
- Increase the SYN backlog queue size.
- Reduce SYN_ACK retries.
- Reduce the SYN Flood attack by lowering the timeout value for SYN_RECV.
- Protect IP Spoofing which is used for SYN Flood attack.

23.3.3 Algorithm for Detecting and Protecting from a SYN Flood Attack

For each packet arrival do
Check the IP Header Length
if IP Header Length = 20 then
if protocol = TCP then
    If ipaddress is available in server side database then
        Packet is Correct.
            Forward the packet to destination.
   Else
     Check for symptoms of a SYN flood attack if not
        a known bad address.
     if SYN FLOOD ATTACK then
     Notify the administrator about SYN Flood
        Attack.
     Identify the source of the packet and deny
        the packet.
     If newsource then
        Store the source of the packet
            (IPaddress)in bad address database.
        End if
     End if

End if
End if
End if
End

23.4 TCP Sequence Number Attack

One end of the TCP session is controlled by the attacker. The attack will be successful when the attacked end of the network is tricked for the duration of the session. In this attack the attacker can hijack or disrupt the session by responding to the sequence number similar to the one used in the original session. If the attacker guesses the valid sequence number then the attacker gains the connection and the data from the legitimate system. Initiating TCP sequence number spoofing through predicting TCP initial sequence number has become a hidden trouble on TCP protocol [9].

The communication session between the target and the trusted host can be exploited by the TCP sequence number.

The attacker can perform TCP Sequence Number attack in the following sequence when Host X and Host Y communicate with each other.

- The attacker wants to attack Host Y.
- It floods Host X with new requests causing a Denial of service attack to stop Host Y from communicating with X.
- The attacker T connects to its target Host Y. It then impersonates X and sends

$$Xt \rightarrow Y : SYN, ISNt$$

Where "Xt" denotes a packet sent by T pretending to be X. It is a Faked packet, Host Y thinks it is coming from Host X.

Y's response to T's original SYN

$$Y \rightarrow X : SYN, ISNy, ACK(ISNt)$$

This faked packet can cause terminating connection or allow the Host Y to perform malicious command/scripts etc.

23.4.1 Defending from a TCP Sequence Number Attack

The best idea is to allow the gateways to reject the external packets into the local net that claim to be from the local net. Block the packets with the internal source addresses arriving on the external interface and to stop the attacks originating from the site, block the packets with a source address different from the internal network. One of the best practices to defend this attack is using TCP stacks with less predictable Initial Sequence Numbers (ISNs).

23.4.2 Algorithm for Defending from TCP Sequence Number Attacks

For each packet arrived at the gateway do
    Check for the incoming interface.
    If source interface is external then
        Check for source address spoofing.
        If source address is internal then
            Drop the packet.
        Else
            Forward the packet.
        End if
    Else
    Source interface is internal, check for spoofed packet.
        If source address is external then
            Drop the packet.
        Else
            Forward the packet.
        End if
    End if
End

23.5 TCP Hijacking

In TCP Hijacking the attacker gains access to a host in the network and logically disconnects it from the network. Another machine is inserted with the same IP address by the attacker. This happens quickly and gives the attacker access to the session and to all the information on the original system. The attacker can hijack the session, observe the sequence numbers and acknowledgement numbers of both the hosts. Then the attacker can send the spoofed packet where the recipient cannot authenticate the source of the packet and accept it as a valid packet. The both hosts will notice the disturbance in the network service as the expected sequence numbers and acknowledge numbers will not match. As a result, both hosts may suspect an attack and consider the packet as invalid packet.

23.5.1 Defending from a TCP Hijacking

If TCP Hijacking occurs than the critical information will be gathered during the attack. TCP Hijacking attacks are performed using Telnet, FTP, DNS and unencrypted protocols. TCP Hijacking attacks are avoided by using encrypted protocols such as SSH, SSL, and IPSec. The session keys generated by encrypted protocols provide secured communication channel and the attacker cannot be able to get session keys. Digital Signature can provide strong protection against the TCP Hijacking, if the attacker steals the session keys.

23.6 ICMP Smurf Attacks

Attacker uses fake IP address to send messages to a computer as if the messages are coming from a trusted host. The main aim of smurfing is to conceal sender's identity. Smurfing makes use of Internet Protocol (IP) and Internet Control Message Protocol (ICMP). Smurf attack floods a system via spoofed broadcast ping messages. The attacker sends large number of ICMP echo requests with spoofed source IP addresses to IP broadcast addresses. Most of the hosts on this IP network will accept the ICMP echo request and reply to it with an echo reply. Due to large amount of echo replies from multiple hosts would consume large amount of network bandwidth and result in slow down of network. Any broadcast enabled network or any host responding to broadcast address can be a potential target for ICMP Smurf attacks. Solutions to the smurf attack include disabling the IP-directed broadcast service at the intermediary network [11].

23.6.1 Defending from ICMP Smurf Attacks

Following are the steps to prevent ICMP Smurf attacks.
- Individual hosts and routers need to be configured not to respond to ping requests or broadcasts.
- Routers need to be configured not to forward packets directed to broadcast addresses.
- Do not allow your firewall to accept ICMP echo requests from the Internet.
- Restrict the flow of information outbound from one network to another to ensure that Smurf attack is not launched.
- Simply block all inbound and outbound ICMP *echo-request* and ICMP *echo-reply* packets.
- Depending upon the configuration forward the packets and log the event or drop the packet if the destination address of a received ICMP echo request message is a subnet broadcast address or network address.

23.6.2 Algorithm for Defending from ICMP Smurf Attacks

For each packet arrived at the router do
  Check for ICMP echo messages
  If type of packet is ICMP echo-request or ICMP
    echo-reply then
      If configuration-rule is to allow then
        Check for destination address.
        If ipdestination is subnet or
          broadcast then
          Drop the packet.
          Log the event.
        Else
          Forward the packet.
End if
      Else
        Drop the packet.
      End if
    Else
      Forward the packet.
    End if
End

23.7 Packet Spoofing

Packet spoofing is a difficult type of attack to tackle. Many different types of attacks can be classified as packet-spoofing attacks [18]. The spoofed packet would contain the seqnum and acknum expected by the recipient and the source address of the other host [19]. The large amount of traffic will effectively spoofed the victim, if there are many hosts in used networks [16]. It can result in serious data failure, and there are ways to detect it and stop it. Finance firms are found to be common victims to this kind of attack. This is a generic type of attack that may be used in any Denial of Service attack.

23.7.1 Algorithm for Defending from Packet Spoofing

For each packet arrived at the router do
  Check for authenticity of the sender.
  Request ACK from the host claiming to be the sender.
  If no ACK received then

>   *Drop the packet and log the action.*
>   *Else*
>   *Forward the packets at the router to the*
>   *hosts in the internal network.*
>   *End if*
> *End*

23.8 Statistical Comparison Between Different Types of DoS Attacks

DoS attacks have no boundaries and have hit all the sectors of the industry such as financial services, banking, insurance, hospitality, travel, government organizations, defence, etc. The attacks mainly focus on bandwidth capacity and routing infrastructure. It has been observed that most used DoS attack is TCP SYN Flood as summarized in Table 23.1.

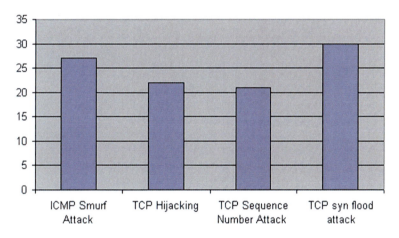

Bar graph representing attacks and their usage

23.9 Safety Measures to Prevent Different Types of DoS Attacks

Key kernel parameters can be adjusted to immunize Linux systems (for instance, in the file/etc/sysctl.conf) from typical DoS attacks to a certain degree as summarized in Table 23.2.

Table 23.1 Statistical comparison between different types of DoS attacks

Type of attack	How attack works	Impact	DoS attacks most used (approximately) (%)
ICMP Smurf attack	Floods a system via spoofed broadcast ping messages	Causes the victim's server to crash	27
TCP Hijacking	Attack may be used to gain illegal access to system resources	All unencrypted TCP protocols are susceptible for this type of attacks	22
TCP Sequence Number attack	Attacks exploit the communication session, which was established between the target and the trusted host that initiated the session	This attack disrupts or hijacks a valid session	21
TCP Syn Flood attack	Exploits weakness in TCP/IP protocol	Affects the resources that run TCP server processes	30

Table 23.2 Safety measures to prevent different types of DoS attacks

Attack	Parameters	Recommended value
TCP SYN Flood attack	Using SYN cookies	Enable
	Increasing the SYN backlog queue	2,048
	Reducing SYN_ACK retries	3
	Setting SYN_RECV timeout	40
	Preventing IP spoofing	Enable
TCP Sequence Number attack	Forwarding of source routed packets	Disable
TCP Hijacking attack	Source ports	32,768–61,000
ICMP Smurf attacks	Ignore Smurf attacks	Enable
	Forward traffic between interfaces	Disable
	Protect against SYN Flood attacks	Enable
	ICMP Redirect Acceptance	Disable
	Burst-normal value	1–100,000
	Burst-max value	1–100,000
	Lockup-value	1–10,000

23.10 Results

Figure 23.1 indicates that the server was attacked by the SYN Flood attack at approximate Time (T) 110 s. It has been observed that the number of packets captured at the network interface of the server machine is more than 1,500 packets per second during the SYN Flood attack. This indicates that the SYN Flood attack consumes network bandwidth and resources on the server machine. When the implementation of algorithm 3.3 is executed at Time (T) 122 s the number of

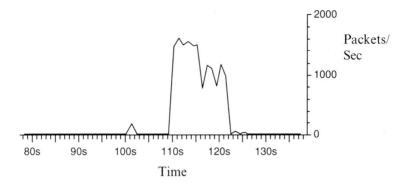

Fig. 23.1 Network traffic observed at the server before and after implementation of defence algorithm during a simulated SYN Flood attack

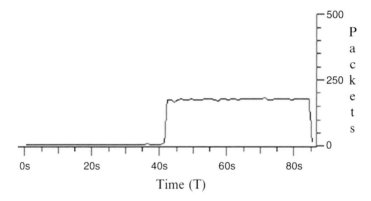

Fig. 23.2 Network traffic observed at the server before implementation of defence algorithm during a simulated ICMP Smurf attack

packets captured by the machine have been dropped to less than 10 packets per second which is in the range of current system normal load.

Figure 23.2 depicts network traffic analysis at the server machine before and during a simulated ICMP Smurf attack. It has been observed that while the average number of packets before the attack occurring at approximate Time (T) 41 s was around 10 packets per second, the number of packets captured at the network interface of the server machine is more than 200 packets per second during the ICMP Smurf attack.

Figure 23.3 depicts network traffic analysis at the server under the simulated ICMP Smurf attack after execution of the implementation of algorithm 6.2. The total number of packets received per second on the server after defending the attack at approximate Time (T) 161 s on the server by applying algorithm 6.2 (i.e. by dropping the packets if the type is ICMP echo request or ICMP echo reply) has stabilized the network traffic back to the earlier observed normal operating load (Figs. 23.4 and 23.5).

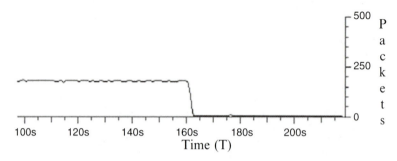

Fig. 23.3 Network traffic observed at the server after implementation of defence algorithm during a simulated ICMP Smurf attack

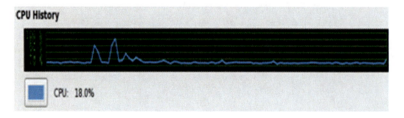

Fig. 23.4 Shows CPU usage on the server when the server was affected by ICMP Smurf attack

Fig. 23.5 Shows CPU usage after defending ICMP Smurf attack on the server

23.11 Conclusions

Global businesses are shifting towards emerging technologies such as service oriented architecture using cloud computing for a sustained revenue model. Ensuring round the clock uninterrupted service to the clients thus, becomes a top priority to any organization. This paper discuses the different types of Denial of Service attacks specific to Linux server platforms and presents ways to detect and prevent such attacks or to avoid such attacks altogether so that the server platforms that form the backbone of service oriented architecture function smoothly without a breakdown.

References

1. A.C. Snoeren, Hash-based IP Traceback, in *Proceedings of the ACM SIGCOMM Conference*, 2001, pp. 3–14
2. B.Q.M. AL-Musawi, Mitigating DoS/DDoS attacks using iptables. Int. J. Eng. Technol. IJET-IJENS **12**(3) (2012)
3. B.B. Gupta, R.C. Joshi, M. Misra, Distributed Denial of Service prevention techniques. Int. J. Comput. Electr. Eng. **2**(2), 268–275 (2010)
4. B. Xiao, W. Chen, Y. He, An autonomous defense against SYN Flooding attacks: detect and throttle attacks at the victim side independently. J. Parallel Distrib. Comput., Elsevier **68**, 456–470 (2007)
5. C. Patrikakis, M. Masikos, O. Zouraraki, Distributed Denial of Service attacks. Internet Protoc. J. **4** (2007). ISSN 1944-1134
6. C.-M. Cheng, H.T. Kung, Koan-Sin Tan, Use of spectral analysis in defense against DoS attacks, in *Proceedings of IEEE GLOBECOM 2002*, vol. 3, November 2002, pp. 2143–2148
7. C.L. Schuba, I.V. Krsul, M.G. Kuhn, E.H. Spafford, A. Sundaram, D. Zamboni, Analysis of a Denial of Service attack on TCP, in *Proceedings of the 1997 I.E. Symposium on Security and Privacy*, pp. 208
8. C.L. Schuba, I.V. Krsul, M.G. Kuhn, E.H. Spafford, A. Sundaram, D. Zamboni, Analysis of a Denial of Service attack on TCP, in *Proceedings of the IEEE Symposium on Security and Privacy*, IEEE Computer Society Press, Silver Spring, MD, 1997, pp. 208–223
9. F. Zeng, Research on TCP Initial Sequence Number prediction method based on adding-weight chaotic time series, in *Proceedings of IEEE, ICYCS 2008*, pp. 1511–1515
10. H. Wang, D. Zhang, K.G. Shin, Detecting SYN Flooding attacks, in *Proceedings of IEEE INFOCOM 2002*, June 2002, pp. 1530–1539
11. J. Sen, A robust mechanism for defending distributed denial of service attacks on web servers. Int. J. Netw. Secur. Appl. **3**(2), 162–179 (2011)
12. K. Kumar, R. Joshi, K. Singh K, An integrated approach for defending against Distributed Denial of Service attacks (2002), http://www.cs.iitm.ernet.in/~iriss06/paper.html
13. L. Yun, G. Ye, W. Guiyi, Detect SYN Flooding attack in edge routers. Int. J. Secur. Appl. **3**, 31–45 (2009)
14. L. Kavisankar, C. Chellapan, Challenging Number Approach for uncovering TCP SYN Flooding using SYN Flooding attack. Int. J. Netw. Secur. Appl. **3**(5), 191–202 (2011)
15. M.E. Manna, A. Amphawan, Review of SYN-Flooding attack detection mechanism. Int. J. Distrib. Parallel Syst. **3**(1), 99–117 (2012)
16. M. Bogdanoski, A. Risteski, Wireless network behavior under ICMP ping flood DoS attack and mitigation techniques. Int. J. Commun. Netw. Info. Secur. **3**(1), 17–24 (2011)
17. M. Voznak, J. Safarik, DoS attacks targeting SIP server and improvements of robustness. Int. J. Math. Comput. Simul. **6**(1), 177–184 (2012)
18. N.A. Noureldien, M.O. Hussein, Block Spoofed Packets at Source (BSPS): a method for detecting and preventing all types of spoofed source IP packets and SYN Flooding packets at source: a theoretical framework. Int. J. Netw. Commun. **2**(3), 33–37 (2012)
19. O. Zheng, J. Poon, K. Beznosov, Application-based TCP Hijacking, in *EUROSEC '09 Proceedings of the Second European Workshop on System Security*, pp. 9–15
20. S. Pukkawanna, V. Visoottiviseth, P. Pongpaibool, Lightweight detection of DoS attack, in *Proceedings of IEEE ICON2007*, Adelaide, South Australia, November 2007
21. T. Anderson, T. Roscoe, D. Wetherall, *Preventing Internet Denial-of-Service with Capabilities*, Intel Research Berkeley, Intel Corporation, Copyright 2003
22. W.M. Eddy, Defenses against TCP SYN Flooding attacks. Internet Protoc. J. **9**(4) (2006)
23. W. Liu, Research on DoS attack and detection programming, in *2009 Third International Symposium on Intelligent Information Technology Application*, vol. 1, IEEE Computer Society Washington, DC, pp. 207–210. ISBN 978-0-7695-3859-4
24. X. Yang, Typical DoS/DDoS threats under IPv6, in *Computing in the Global Information Technology, ICCGI 2007*, pp. 55. ISBN 0-7695-2798-1
25. Y. Chen, S. Das, P. Dhar, A. El Saddik, A. Nayak, Detecting and preventing IP-spoofed distributed DoS attacks. Int. J. Netw. Secur. **7**(1), 70–81 (2008)

Part V
Wireless and Mobile Networks

Chapter 24
Cache Coherency Algorithm to Optimize Bandwidth in Mobile Networks

Abhinandan Ramaprasath, Karthik Hariharan, and Anand Srinivasan

Abstract Mobile networks are becoming popular in catering to Internet through smart phones. The next generation phones provide ubiquitous communication of voice, video and data through the hand held devices. Unfortunately, the increase in service provider capacity has not kept up with the user demand for more bandwidth. It is becoming very expensive for service providers to cater higher bandwidth without investing on new technology or expansion. In this paper we propose a bandwidth optimization algorithm based on cache coherency where the user data transfer is optimized without compromising the user expectation or the need for service providers to expand their capacity. The proposed algorithm is compared with existing data transfer techniques and we show through representative analysis the efficiency of the algorithm to keep the same level of communication with less transfer. To emphasize the practicality of the algorithm, we also provide some insights into how it is implemented.

24.1 Introduction

With the growing use of mobile phones throughout the world, the Internet is becoming increasingly prevalent on mobile phones. Mobile Internet as it is popularly called these days' enables users to access the internet on the go. While there are many advancements being made in communication speeds through technology and new standards, service providers are unable to invest at the same speed. This has led the service providers to cater to new wave of mobile Internet users with existing infrastructure.

A. Ramaprasath (✉) • K. Hariharan
Department of Information Technology, SSN College of Engineering, Chennai, India
e-mail: abhiin1947@gmail.com; karthikhariharan13@gmail.com

A. Srinivasan
Department of System and Computer Engineering, Carleton University
and EION Inc., Ottawa, Canada

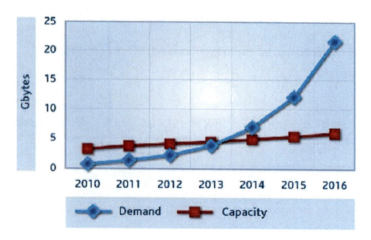

Fig. 24.1 Demand, time plot and capacity, time plot [2]

According to studies done by 4G Americas [1] and several white papers published the capacity of the networks scales linearly, while the demand seems to increase exponentially. The graph below from the 4G Americas shows that the trend is going to continue and only gets amplified (Fig. 24.1).

In most developing countries, mobile phone popularity and usage is growing at an increasing rate. In such countries, there are more and more people accessing content on the Internet, and bandwidth is of utmost importance. Bandwidth can be optimized by either increasing the capacity of the spectrum or increasing data transmission rates. Unfortunately, both these solutions lead to service providers spending more money for infrastructure leading to pressure on consumers to pay more. However, there is another possible solution that could optimize the bandwidth to customers without compromising the performance or increasing expense. In this paper we explore the way to improve the bandwidth per user within the same infrastructure by proposing a novel method.

24.2 Proposed Solution

Presently, when a mobile phone accesses the Internet, the inherent protocol transfers the entire data back to the phone. The data transferred from the base station to the end device (i.e., phone) occupies the most important resource in the air, namely the bandwidth.

The current algorithm is as follows:

```
READ requestedFile, cachedFiles
IF requestedFile NOT IN cachedFiles THEN
    READ fetchedFile FROM networkLoc
    ADD to cachedFiles
ELSE
    IF cachedFiles NOT up to date THEN
        READ fetchedFile FROM networkLoc
        REPLACE outdatedFile IN cachedFiles
    ELSE
        READ requestedFile FROM cachedFiles
    ENDIF
ENDIF
```

Whenever a file is requested, the cache is checked for the consistency and availability of the requested data. If the data is not present, then the file is fetched from the network location and the cache is updated with the requested file.

The algorithm looks simple enough to implement, but we argue the inherent inefficiency in using such an algorithm for a mobile device communication. It may make sense when a person uses a fiber link to upload the file to cache, however to download the requested file blindly bleeds the most expensive resource, namely the bandwidth in a mobile network.

We propose a solution that is simple to implement, practical and optimizes the network bandwidth in mobile networks. Our proposed bandwidth optimization algorithm is:

```
READ requestedFile, cachedFiles
IF requestedFile NOT IN cachedFiles THEN
    CLONE cache GIT repository
    ADD to cachedFiles
ELSE
    IF cachedFile NOT up to date THEN
        PULL from cache git repository
        STORE in cachedFiles
    ELSE
        READ requestedFile FROM cachedFiles
    ENDIF
ENDIF
```

The solution being proposed uses GIT subversion control system [3]. GIT subversion control system [4] operates with a high speed, small in size to implement on a machine, its high levels of compression [5] and its distributed nature is very suitable for optimization. We bring in the subversion concept used in general non networking versioning system, to mobile communication world for the first time and show that it can be used for optimizing network performance.

24.2.1 Better Use of Bandwidth

The bandwidth optimization algorithm achieves high speed of transfer that is better than any other known systems. One of the ways that this is achieved is by calculating the differences in the old source file and the new source file that is being uploaded. It gets only the changed lines and the line before and the line after the corresponding line that has to be changed. The changes are only made to those corresponding lines instead of overwriting the entire file thus leading to optimized transfer of data. Only the needed and changed data is transferred and not the whole file.

The cache coherence in the algorithm being proposed in this paper minimizes the data transfer without compromising the customer expectation. Website developers and vendors will maintain GIT repositories of the most frequently downloaded files of that website. A version number is maintained on both the client end and the server end. If the version numbers match, then the cache is up to date and the local version can be used. In case the version numbers do not match, using the technique mentioned above, the changes in the files are downloaded and merged with the existing versions of the cache. This will involve in far less data being transferred over the wire and hence better utilization of bandwidth.

24.2.2 Small Size of Repositories

If cache coherency were to be achieved using the GIT subversion system, then there will be a necessity for GIT repositories on the client's cache. However, the size of the GIT repository in terms of its configuration files and structure files is minimal and small. This will lead to more space for actual files that are to be cached leading to space optimization.

24.2.3 Distributed Nature

Not all subversion systems have been distributed. This is an essential feature for our cache coherence system, as the server and its many clients are systems that are completely on either ends of a connection. This distributed nature is what will allow the cache on the client side to check the version and download the corresponding changes that are to be fetched (if any). This results in the data consistency.

24.2.4 Compression

GIT compresses files using the famous zlib library that compresses files to smaller sizes. The zlib library uses a combination of the LZ77 algorithm and Huffman coding called DEFLATE [6]. It maintains a dictionary of all the strings that have been encountered. If a word that is encountered is already present in the dictionary, then it places the index of the entry in the compressed string. If the string doesn't exist then it is appended into the dictionary. The DEFLATE algorithm gives a lot of flexibility and importance can be given to speed of encoding or level of compression. GIT focuses on the compression level which greatly reduces the size of the data that is transmitted over the network.

24.3 Performance Analysis

We assume the following for analysis:

1. 2G data is based on either GPRS [7] or EDGE [8] technologies. This is common and over 80 % of the world including India follows this technology.
2. We consider the file bootstrap.css from the famous UI library bootstrap. This clearly reflects the normal bootstrap mechanism during data transfer.
3. This analysis only considers the time taken to fetch the file into the cache as that impacts the bandwidth and resource estimation.
4. No client or server side processing times are included in this analysis since the goal of the performance is to optimize the bandwidth. Largely the processing time is dependent on the processing capabilities of the end units.

We analyze by using a boot strap of the size approximately 826 Kbits, where a color is changed from red to green.

Size of bootstrap.css: 826,512 bits
if the line

```
        ...
     body {
       margin: 0;
       font-family: "Helvetica Neue", Helvetica, Arial, sans-serif;
       font-size: 14px;
       line-height: 20px;
       color: #333333;
       background-color: #ffffff;
     }
        ...
```
is changed to

```
...
body {
  margin: 0;
  font-family: "Helvetica Neue", Helvetica, Arial,   sans-serif;
  font-size: 14px;
  line-height: 20px;
  color: #ffffff;
  background-color: #333333;
}
```

Let us check on how this impacts the resource utilization in different technologies if current algorithm is used, and how it changes when our proposed algorithm is used.

24.3.1 GPRS

Bandwidth of the link is 56 Kbits/s.
Download speed: $56/8 = 7$ Kbits/s.

24.3.1.1 Performance of Current Algorithm

The entire file bootstrap.css is downloaded (826,512 bits) and the time required to download: $993,784/7,000 = 141.969$ s.

24.3.1.2 Performance of Proposed Algorithm

We calculate the difference between what is asked and what is in cache. The diff file generated by git diff is of size: 2,960 bits.

Time required for downloading: $2,960/7,000 = 0.423$ s.

% increase in performance $= 141.969/0.423 * 100 = $ **33562.411** % which is representative.

It is to be noted that most of the data intensive transfers such as video are very prone to incremental changes and do not require large scale downloading of entire files. The incremental changes, namely diff, are appropriate for download thus optimizing network performance. Smaller the difference, more efficient is the proposed algorithm.

24.3.2 EDGE

Bandwidth of the link is 236.8 Kbits/s.
Download speed: 236.8/8 = 29.6Kbits/s.

24.3.2.1 Performance of Current Algorithm

The entire file bootstrap.css is downloaded (826,512 bits) Time required to download is 993,784/29,600 = 33.574 s.

24.3.2.2 Performance of Proposed Algorithm

The diff file generated is of size: 2,960 bits.
Time required to download is 2,960/296,000 = 0.01 s. The percentage of increase in performance is 33.574/0.01 * 100 = **335,740 %,** which is representative again.

The performance of the proposed algorithm, in the worst case, would be same as the current algorithm. On an average, we will have a great improvement over current algorithm, when proposed algorithm is used as the mobile applications have incremental data to transfer.

24.4 Bandwidth Optimization Algorithms: Data Transfer Mechanisms

The proposed algorithm works by calculating the difference between two commits based on GIT. This is done row-wise in the case of text files and bit difference in the case of binary files. The data is stored as database files in the folder ".git" present in the root of the repository. Once the difference file is downloaded from the internet it is first decompressed and then the difference algorithm is implemented on the database with the help of the difference file. This is again stored in a new database. Difference file contains initial header data and for each changed line, the line number, previous line and changed line are stored. To decrease the size of the difference file in the case of minified files, binary and text difference can both be calculated and the smaller one may be uploaded.

GIT transfers data using one out of four existing protocols – http/https, git, ssh and local. Since all browsers will support the http and https, communication is not worry. The http/https protocol was revised to work just as efficiently as that of git or ssh starting from version 1.6.6 and hence there is no concern about the efficiency of the http/https protocol.

24.5 Implementation Details

The implementation is on both sides, namely the server and the client. The implementation on the client side included a GIT client available in the client's device or integrating it with the browser. When properly managed, the same repository could also be used to cross-browse, because of the fact that many mobile users tend to have more than one browser installed on their devices. The GIT repositories are stored in an allocated folder with a max size cap. The client side's GIT repository is a minimal version stripped of most of the functionality of a traditional GIT client. This is because the client is only required to download the difference file and then change the source file. The browser only clones the repositories of the websites that it thinks the user visits regularly. This is implemented using a simple most frequently used algorithm. If the size of folder increases beyond the max cap then the browser deletes the repository of the website that is least frequently used. The file system used here could be a persistent HTML5 file system.

The implementation on the server side required a GIT server. The GIT repository resides on the server under the path "/git/cache.git". For example, the repository for the website "http://www.google.com/" would be in "http://www.google.com/git/cache.git". If the server does not prefer this path a new path is specified in the page within a "meta" or "git" tag. The server stores only the most used css and js files.

24.6 Conclusions and Future Work

We introduced a bandwidth optimization algorithm based on cache coherency and GIT technique towards optimizing bandwidth in a mobile network. This allows the service providers to support higher data transfers using existing networks and future data intensive mobile handsets. We also showed through analysis the efficiency of the proposed algorithm. We have implemented a prototype and it is being tested.

This algorithm works very well in popular websites such as Facebook and Gmail that never drastically change their user interface. It is easy to cache or clone such data for faster response. Our work in future is to find an optimal way to populate GIT repository for esoteric websites that people do not visit often. One method is to have a hybrid mechanism that combines two techniques based on hits. Another interesting field is to research the impact of data tampering and protection/encryption mechanisms for data consistency and accuracy.

References

1. 4G Americas, 4G mobile broadband evolution – Series of white papers, http://www.4gamericas.org
2. Rysavy Research, Mobile broadband capacity constraints and the need for optimization, http://www.rysavy.com/Articles/2010_02_Rysavy_Mobile_Broadband_Capacity_Constraints.pdf
3. J.C. Hamano, GIT – a stupid content tracker, in *Proceedings of the Linux Symposium 2006*, Ottawa, Canada, 2006
4. S. Chacon, Pro Git, Apress publications, Section 4.1 to 4.11 and Section 9.2 to 9.6
5. J.-L. Gailly, M. Adler, Compression algorithm (deflate), http://www.gzip.org/algorithm.txt
6. J. Ziv, A. Lempel, A universal algorithm for sequential data compression. IEEE Trans. Info. Theory **23**(3), 337–343 (1977)
7. Wikipedia, General Packet Radio Service, http://en.wikipedia.org/wiki/GPRS
8. Wikipedia, Enhanced data rates for GSM evolution, http://en.wikipedia.org/wiki/EDGE

Chapter 25
Connected Dominating Set Based Scheduling for Publish-Subscribe Scenarios in WSN

P. Kaviya and R. Ramalakshmi

Abstract An important issue in Wireless Sensor Network(WSN) is to prolong network lifetime. Sensors are hard to recharge since it is battery powered. Network lifetime can be increased by optimizing the energy consumption. Sensor scheduling is a mechanism used to maximize the network life time. One method to conserve energy is to put sensors in sleep mode when they are not actively participating in sensing and data forwarding. In publish-subscribe scenarios, the sink sends a query as interest toward sensors in the moritored area and all the sensors perform the sensing task if they can satisfy the interest. In this paper, we propose to select a set of active sensors called Connected Dominating Set (CDS), to perform sensing and data transferring. The remaining sensors will be in sleep state to save energy. In this approach, nodes with high residual energy and degree of reachability are selected to form the CDS. The sensors change their role alternatively to prolong network lifetime. We evaluated the performance of the proposed work in terms of number of active sensors, total energy consumption, delivery ratio and delay.

25.1 Introduction

Sensors are used to monitor and control the physical environment [1]. A Wireless Sensor Network consists of large number of sensor nodes that are densely deployed. Sensor nodes measure various parameters of the environment and transmit the

P. Kaviya
Department of Computer Science and Engineering, Sri Vidya College
of Engineering & Technology, Virudhunagar, TN, India
e-mail: kaviyaap@gmail.com

R. Ramalakshmi (✉)
Department of Computer Science and Engineering, Kalasalingam University,
Krishnankoil, Srivilliputtur, TN, India
e-mail: ramalakshmir@yahoo.com; rama@klu.ac.in

collected data to the sink using hop by hop communication. The sink processes the sensed data and forwards it to the users. Energy management is an important issue in WSNs. The major component that consumes energy is radio, which can be in one of the following modes: sleep, idle, transmit, and receive. When the host is not transmitting or receiving data means a radio is in idle state, and usually at the receive mode the power consumption is high. Both the transmitter and the receiver are turned off, when a radio is in sleep mode which is the most energy efficient state.

In publish-subscribe scenarios, the sink sends queries to the monitor area from time to time. Sensors are equipped with one or more sensing components. When sensors receive a query, they are activated to perform sensing task, only if they satisfy the requirements that specified in the query. The sensed data are transmitted to the sink through intermediate sensors.

Sensor scheduling mechanism serves different types of queries. To prolong network lifetime, it is not necessary that all sensors remain active all the time and some sensors can go to sleep [2]. We proposed a connected dominating set (CDS) based localized mechanism. Initially, a basic backbone (CDS) is constructed. When a query is issued, new sensors are activated locally such that to meet the query, coverage and connectivity requirements. Some sensors return to sleep state and the CDS backbone remains active, when the query expired. CDS is an energy-efficient mechanism which constructs a backbone in order to route the queries and sensed data. The coverage requirement requires that the related sensors in the monitored area are activated to perform sensing tasks when the queries are propagated to that particular monitored area. This is an adaptive mechanism, new sensors might be activated when query arrives, and sensors in the monitored area go to sleep after the query expires. The connectivity requirement requires that all the time the set of active sensors to be connected and it is necessary for some operations like data reporting, query processing, and forwarding of control messages [3].

The rest of the paper is organized as follows. We discuss previous works on related topics in Sect. 25.2. Section 25.3 describes CDS based Sensor Scheduling in detail. Section 25.4 provides the simulation results of the proposed sensor scheduling mechanism. Section 25.5 finally concludes the paper.

25.2 Related Works

The most important features of a WSN is energy efficiency, as sensors have small batteries that are impossible or impractical to change and recharge. Intanagonwiwat et al. [4] have described about the publish-subscribe scenario. In such scenarios, the sink sends queries as interest toward sensors in the monitored area. When sensors receive a query, if they can satisfy the interests and serve the query, then they are activated to perform sensing task. The intermediate sensors help to transfer sensed data towards the sink. Vu et al. [5] have proposed an algorithm to construct a set of tree-structured detection sets to achieve energy efficient and reliable surveillance. In order to achieve reliable surveillance, each atomic event part of the composite event must be watched by at least k sensors. The algorithm has designed to work in greedy manner. Wu et al. [6] have presented a localized solution for building a CDS in ad hoc wireless networks. They proposed a CDS based backbone which provides an efficient mechanism to achieve global connectivity using a localized mechanism. Also they proposed a dominant pruning rule to reduce the size of the dominating set. Yang et al. [7] have designed an adaptive sensor scheduling mechanism, in which sets of active sensors are chosen to work alternatively in order to serve different types of queries, to achieve global connectivity requirements, and to maximize the network lifetime. Rakhi Khedikar et al. [8] have discussed the various methods of scheduling for increasing the lifetime of the network by reducing the energy consumption. Optimal scheduling algorithm for sensors to enhance the lifetime of wireless sensor networks are developed.

Rachid Saadi et al. [9] have discussed an adaptive scheduling of sensor node's activity to extend the network lifetime. A distributed algorithm is proposed that ensured both coverage of the deployment area and network connectivity by providing multiple set covers. Different coverage formulations have been proposed by Mihaela Cardie et al. [10] and discussed about various coverage approaches with random sensor deployment. Stojmenovic et al. [11] have proposed new metrics for source – independent localized dominating set, based on combinations of node degrees and remaining energy levels, for deciding activity status. Also they proposed CDS scheme to prolong network lifetime and to preserve global connectivity. Kumar et al. [12] have described that sensor nodes may be equipped with different numbers and types of sensing components due to the following reasons: they might be manufactured with different sensing capabilities, some of the sensing components of a sensor may lack of memory for storing data, or some sensor components might fail over time. Torkestani [13] has concentrated on backbone lifetime, transmission delay, backbone size, and control message overhead. The delay-bound energy-efficient backbone formation problem in WSN have been modeled as the degree-constrained minimum weight connected dominating set (CDS) problem, where the residual energy is assumed as the node weight. Mihaela Cardie et al. [14] have proposed an efficient method to extend the sensor network lifetime by organizing the sensors into a maximal number of set covers that are activated successfully. Only the sensors from the current active set are responsible for monitoring all targets and for transmitting the collected data, while all other nodes are remain in sleep mode. Razieh Asgarnezhad et al. [15] have described about various CDS backbone formation algorithms. The CDS backbone reduces communication overhead and overall energy consumption, increases bandwidth efficiency and network lifetime in a WSN.

25.3 Proposed Work

We assumed a wireless sensor network where the sink sends query from time to time and sensors are equipped with one or more sensing components. The proposed work is to design an adaptive sensor scheduling mechanism such that a set of active sensors are chosen to work alternatively to serve different types of queries, to achieve connectivity requirements, and to maximize the network lifetime. A dominating set is a subset of sensors in which every sensor is either in the subset or the subset having its neighbors. The sensors in the dominating set be connected are called Connected Dominating Set (CDS).

Initially, a CDS backbone is constructed. When a query is issued, new sensors are activated locally to meet the requirements of the query and connectivity. The sink sends a query towards sensors in the monitored area. When sensors receive a query, if they can satisfy the requirements specified in the query, then they activate to perform the sensing task. The intermediate sensors route sensed data towards the sink and the results are finally reported to the sink. A CDS requires that the sensors in the dominating set be connected. The active nodes form a CDS at each time to satisfy the coverage requirements. Our main objective of this work is to design a backbone that change over time as new query request arrives or expires.

25.3.1 Background

Definition 1 (Dominating Set). Consider G=(V,E) is a subset D of V such that every vertex not in D is joined to at least one member of D by some edge.

Definition 2 (Connected Dominating set). In a given graph, subset of nodes such that it forms a dominating set in the graph and the sub graph induced is connected. It involves two properties,

- Any node in D can reach any other node in D by a path that stays entirely within D. That is, D induces a connected sub graph of G.
- Every vertex in G either belongs to D or is adjacent to a vertex in D. It implies that D is a dominating set of G. Figure 25.1 shows the CDS.

25.3.2 Sensor Scheduling in WSN

Sensor scheduling mechanism is used to prolong network lifetime. Sensor scheduling mechanism chooses a set of active sensors to perform sensing and data forwarding. All other sensors go to sleep state to save energy. After some time period, another set of active sensors is chosen. Thus sensors work alternatively to maximize network lifetime. In this work, the active nodes form a CDS at each time to satisfy the coverage requirements. All sensors run CDS algorithm to choose a starting backbone. To forward data or any control messages, CDS nodes are used. When a new query regarding sensing a specified rectangular area is received/expired, sensors adaptively update the active nodes in that area for both coverage requirements and global connectivity. The main operations are CDS formation and Sensor Scheduling.

25 Connected Dominating Set Based Scheduling for Publish-Subscribe Scenarios... 311

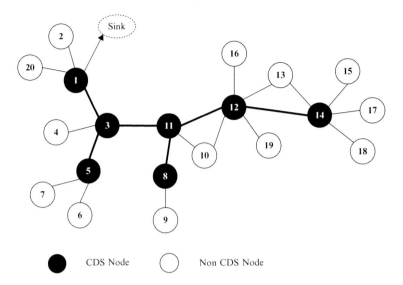

Fig. 25.1 Connected dominating set

Table 25.1 Notations

Variables	Descriptions
t(u)	Amount of time the sensor node u has to be active
p(u)	Priority p (u) = {D (u), E (u)}
s(u)	Status of sensor node u (alive/sleep)
r(u)	Role of sensors in CDS (active/inactive)
D(u)	Number of one hop neighbors of u
E(u)	Residual energy at u
N(u)	One-hop neighbors of u

The notations used in our algorithm are given in Table 25.1. Initially CDS backbone is selected. The requests can be propagated to the monitored area along the backbone. The sensors that are not included in the active node set, go to sleep state.

25.3.2.1 Active Node Selection (CDS)

Every node learns about its neighbors through hello message transmission. Initially, status of a node is set to non-dominating. After hello message transmission, every node calculates its priority with degree or energy value. A one-hop neighbor with highest energy is marked as dominating node. A node with maximum degree is marked as dominating if the energy values are same. The algorithm for active node selection is presented in Algorithm 1.

Algorithm 1: Active node selection algorithm.
```
1:   for each sensor u do
2:      for all v ∈ N(u) do
3:         if E(u) >E(v) then
4:            CDS = CDS U {u} ;
5:         else
6:            if E(u) == E(v) and D(u) > D(v) then
7:               CDS = CDS U {u}
8:            else
9:               CDS = CDS U {v}
10:           end if
11:        end if
12:     end for
13:  end for
```

25.3.2.2 CDS Scheduling Mechanisms for Serving Queries

A query Q is sent by the sink and has the format, **Query**(*type, interval, area R, Tstart, Tend*). The type field specifies the attributes to be sensed and reported to the sink. The interval field indicates how often data from the monitored area have to be reported to the sink. Area represents the query's area specified using rectangular coordinates. The attributes (e.g. Temperature, Humidity) have to be monitored between T start and T end. The sink sends the query Q to the area R by flooding methods. CDS nodes forward the query information by awake the 1-hop sleeping nodes in the area R. Once the Query message is received, the active nodes in the CDS awake their sleeping nodes in area R. Awaken node checks whether it has a sensing components that is specified in the query's type field, then it become active and perform sensing operation. Finally the data are reported to the sink according to the interval specified in the query. When the query ends, based on the residual energy CDS nodes are updated.

Algorithm 2: Sensor scheduling algorithm.
```
1:   for each sensor u do
2:      // When Query Expires
3:      if u is serving another query then
4:         s(u) = Alive
5:      else
6:         s(u) = Sleep
7:      end if
8:   end for
9:   // At every time interval T
10:  for Time Interval T on Sensor u do
11:     Exchange Hello Messages
12:     Compute E(u) and D(u);
```

13:    $s(u) = Alive$
14:    Update $r(u), p(u)$
15:    // **Form new CDS**
16:    Execute Algorithm 1.
17: **end for**

The reporting sensors check their role field, when the query expires. A sensor is remaining in active state if it is a part of the CDS or serving another query. Otherwise the sensor returns to sleep. To remove the expired query, the node updates its status, time, and role fields. After a period of time, some sensors may run out of energy which are present in the CDS. The CDS is updated every time interval T in order to balance the energy consumption. The sleeping nodes in the monitoring area awake and update their priority based on the residual energy, $p(u) = (E(u), D(u))$ for a node u. Then by exchanging Hello messages, they update their neighbors. The CDS backbone Algorithm 1 is executed to decide the new CDS for the next period T. All the nodes in the network actively participate in the new backbone selection and update their fields based on whether they will be part of the CDS backbone or not. The nodes serving queries will be in active state to perform their sensing tasks. Nodes are return to sleep state which are not serving queries and not in the CDS. The scheduling mechanisum is presented in Algorithm 2. The CDS adaptive scheduling mechanism has several advantages. It provides an energy-efficient connected backbone, to transfer data, control messages and queries among sensors and sink. The CDS nodes can easily awake their sleeping neighbors, when queries arrives to the monitored area. An important issue in balancing the energy consumption is by rotating the nodes in the CDS. This process is given in Fig. 25.2.

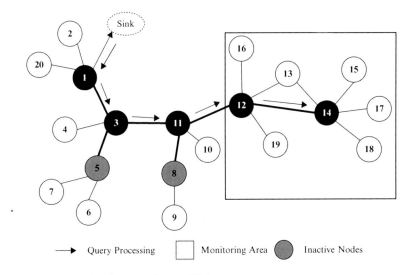

Fig. 25.2 Publish-subscribe scenario over CDS

25.4 Simulation Study

We evaluated the performance of CDS based sensor scheduling algorithm in network simulator ns-2.35 [16]. We placed N sensors, randomly in a field of 200×200 m size. The sensing and communication range are set to 30 m. We study the performance in terms of average number of active sensors, total energy consumption, average end-to-end delay and packet delivery ratio. The network setup is shown in the following Table 25.2.

Average Number of Active Sensors It is the fraction of nodes of the network in the CDS. The degree-based algorithm is mainly focused on node's degree (one hop neighbor) and residual energy of the node. The result of the simulation for a network with 50, 100, 150, 200, 250, 300 and 350 nodes is presented in Fig. 25.3.

Table 25.2 Simulation parameters

Parameter	Values
Area size	200×200 m^2
Channel type	Wireless channel radio
Propagation model	Two ray ground
Sensing and communication range	30 m
Simulation time	600 s
Mac protocol	802.15.4
Sensing interval	5 s
Initial energy	20 J
Transmission power	24 mW
Reception power	14 mW
Packet size	100 bytes

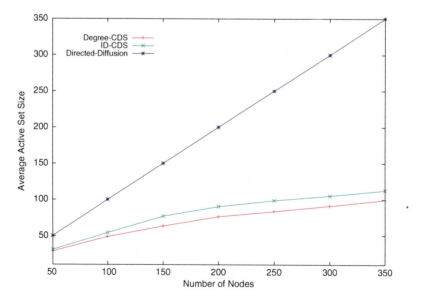

Fig. 25.3 Number of active sensors

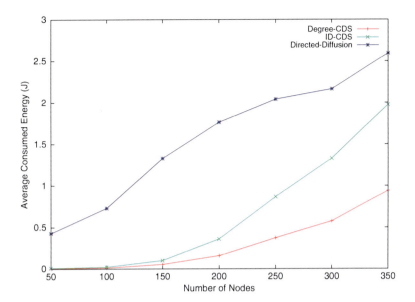

Fig. 25.4 Average consumed energy

It shows that our algorithm has fewer number of active sensors than other algorithms. Including node's degree into CDS selection reduces the CDS size as it covers more nodes. Thus the number nodes included in query processing is minimum and it increases the lifetime of the network.

Average Consumed Energy Figure 25.4 compares the average dissipated energy of the proposed work against existing work and Directed Diffusion. The proposed work consumes less energy for delivering data messages compared with others. In Directed Diffusion, more energy is consumed in transmitting and receiving control messages and more sensors are in the idle state, which consume considerable energy. In CDS based scheduling, only CDS nodes are doing transmission of data and control packets and other nodes will be in sleep state. The number of active nodes in proposed work is less than existing work. Thus the energy consumption is also less in our approach than other algorithms.

Average End to End Delay We measured the delivery time for query messages, data messages, and control messages over the CDS backbone with the scheduling algorithm. The result of the simulation is presented for a network with 50, 100, 150, 200, 250, 300 and 350 nodes. In the Directed Diffusion all sensors are active, so it has shortest data delivery path. Figure 25.5 shows that the proposed work has shorter delivery path when compared to ID-based CDS. The average routing path length from source to sink over minimum size CDS is lower, which has an impact in lesser delay than existing work and Directed Diffusion.

Packet Delivery Ratio It is the successful delivery of data specified in the query. We measured the number of successfully delivered data for the total data over the CDS backbone for the time interval with 2, 4, 6, 8 and 10 s. When the time interval is large, number of data packets to be transfered is less than the low interval time.

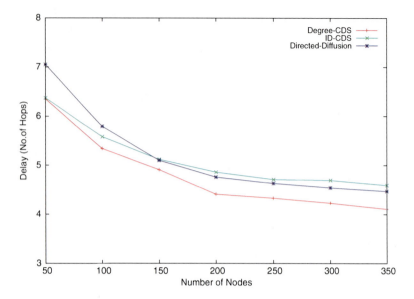

Fig. 25.5 Average end-to-end delay

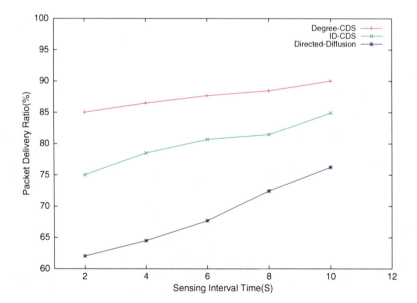

Fig. 25.6 Average packet delivery ratio

Thus the delivery ratio increases with increasing time interval in all mechanisms. The simulation result in Fig. 25.6 shows the average delivery ratio of the proposed work is high than others. This is because only few nodes are active and the active nodes are changing over time.

25.5 Conclusion

The sink sends a query towards sensors in the monitored area. When sensors receive a query, if they can satisfy the interests and query, then they become active state to perform the sensing task. The intermediate sensors forward the sensed data to the sink. Sensor scheduling algorithm is designed which allows sensors not participating in sensing and data transferring to go to sleep in order to conserve energy and to prolong the network lifetime. In this work, we proposed an adaptive energy efficient sensor scheduling mechanism to choose sets of active sensors to work alternatively. A localized CDS selection algorithm is proposed with energy high nodes. We evaluated the performance of the proposed work in terms of energy consumption, delay, delivery ratio and backbone size. The performance of the proposed work is good when compared with existing algorithm which is ID based and Directed Diffusion.

References

1. J. Ben-Othman, K. Bessaoud, A. Bui, L. Pilard, Self-stabilizing algorithm for efficient topology control in wireless sensor networks. J. Comput. Sci. **4**(4), 199–208 (2013)
2. Z. Liu, B. Wang, L. Guo, A survey on connected dominating set construction algorithm for wireless sensor networks. Inf. Technol. J. **9**(6), 1081–1092 (2010)
3. A.G. Ruzzelli, M.J. O'Grady, G.M.P. O'Hare, R. Tynan, Adaptive scheduling in wireless sensor networks, in *Lecture Notes in Computer Science*, **3854**, 266–276 (2006)
4. C. Intanagonwiwat, R. Govindan, D. Estrin, Directed diffusion: a scalable and robust communication paradigm for sensor networks, in *MOBICOM* (ACM, New York, 2000) pp. 56–67
5. C.T. Vu, R.A. Beyah, Y. Li, Composite event detection in wireless sensor networks, in *26th IEEE IPCCC*, New Orleans, Jan 2007
6. J. Wu, H. Li, An extended localized algorithm for connected dominating set formation in ad hoc wireless networks. IEEE Trans. Parallel Distrib. Syst. **15**(10), 908–920 (2004)
7. Y. Yang, M. Cardie, Adaptive energy efficient sensor scheduling for wireless sensor networks. Optim. Lett. **4**, 359–369 (2010). doi:10.1007/s11590-009-01554
8. R. Khedikar, A. Kapur, Y. Survanshi, Maximizing a lifetime of wireless sensor network by scheduling. Int. J. Comput. Sci. Telecommun. **8**(2), 1–6 (2011)
9. A. Makhoul, R. Saadi, C. Pham, Coverage and adaptive scheduling algorithms for criticality management on video wireless sensor networks, in *IEEE, International Conference, ICUMT*, St. Petersburg, 2009
10. M. Cardie, J. Wu, Energy-efficient coverage problems in wireless ad-hoc networks. Comput. Commun. **29**, 413–420 (2006)
11. J.A. Shaikh, J. Solano, I. Stojmenovic, J. Wu, New metrics for dominating set based energy efficient activity scheduling in ad hoc networks, in *Proceedings of the 28th Annual IEEE International Conference on Local Computer Networks*, Bonn/Königswinter, 2003
12. A.V.U.P. Kumar, V.A.M. Reddy, D. Janakiram, Distributed collaboration for event detection in wireless sensor networks, in MPAC'05, Grenoble, 2005
13. J.A. Torkestani, Backbone formation in wireless sensor networks. Sens. Actuator A: Phys. **185**, 117–126 (2012)

14. M. Cardie, M.T. Thai, Y. Li, W. Wu, Energy-efficient target coverage in wireless sensor networks, in *Proceedings of 24th Annual Joint Conference of the IEEE Computer and Communications Societies (INFOCOM 2005)*, **3**, 1976–1984 (2005)
15. R. Asgarnezhad, J.A. Torkestani, Connected dominating set problem and its application to wireless sensor networks, in *The First International Conference on Advanced Communications and Computation, INFOCOMP*, 2011
16. K. Fall, K. Varadhan, The ns Manual, www.isi.edu/nsnam/ns/doc/ns_doc.pdf, 2011

Chapter 26
Ubiquitous Compaction Monitoring Interface for Soil Compactor: A Web Based Approach

R. Prakash, K. Suresh, S. Mydhile Shanmugam, and C.P. Koushik

Abstract Pervasive computing provides an attractive vision for the future. Mobile and stationary devices will dynamically connected and coordinate to seamlessly help people to implement their tasks. However, in reality still there are practices without pervasive monitoring especially in the domain of geo vehicle monitoring. To make the vision of pervasive computing technologies become a constant adaptable to a highly dynamic computing environment we are integrating technologies like GPS (Global Positioning Technology), GSM (Global System for Mobile) and Web services for remote compaction monitoring. GPS is recently being used for wide applications like orbit identification and positioning. GPS need some compatible receivers which support location-awareness using positioning technique like LBS (Location Based System). Soil compaction is a form of physical degradation resulting in densification and distortion of the soil where biological activity, porosity and permeability were reduced, soil strength is increased and soil structure partly destroyed. Monitoring the soil compaction manually in the workspace is not reliable and could not monitor continuously by a single person. This would be the motivation to choose wireless communication. The compaction data and the location data will be sent to the server for remote monitoring. GSM will allow us to transmit the data to the remote server. The objective of this paper is to provide better accuracy with low cost GPS receiver's positioning results. This paper makes use of GPS, ARM7/TDMI (LPC2378) family and GSM. NMEA (National Marine Electronics Association) data acquisition from GPS is monitored. Compaction input is interfaced with GPS co-ordinates. Alerts can be sent from the vehicle to the user mobile phone through GSM communication using AT commands. The remote server should be capable of accepting multiple connections at the same time.

R. Prakash (✉) • K. Suresh • S. Mydhile Shanmugam • C.P. Koushik
P.G. Students, Velammal Engineering College, Chennai, India
e-mail: prakash.rama121@gmail.com; ksuresh0804@gmail.com; mydhile.infotech@gmail.com; cpkoushik@gmail.com

26.1 Introduction

Accessibility of Internet over the GPS platform to nearly 700 Million mobile users across India has created immense opportunities for business to business and business to consumer – marketing, retailing, geo-facility mapping, product/service enquiries, health care delivery etc. As for remote data communication and control, wireless modem was a popular choice. Lots of wireless communication technologies exist that can transmit the data into a particular range, but for remote monitoring from any place we have the data to be globalised. For this globalised data transmission the GSM technology was very famous and the data transmitted from this technology will be globalised and can access it from anywhere in the world.

26.1.1 Pervasive Computing

Pervasive Computing aids Omni connectivity. The idea behind the pervasive computing in information technology system is to manage information easily. In other words accessibility of data is increased. The features of this computing makes connectivity among technologies like GPS, GSM etc. Integrating these technologies with compaction monitoring system makes the system adaptable for soil compactor.

26.1.2 Overview About Soil Compaction

Soil compaction is the process in which a stress applied to a soil causes densification as air is displaced from the pores between the soil grains. When stress is applied the causes densification due to being displaced from between the soil grains then consolidation, not compaction, has occurred [4] . The selection of the most suitable method depends on a variety of factors, like, soil conditions, required degree of compaction, type of structure to be supported, maximum depth of the compaction, as well as site-specific considerations such as sensitivity of adjacent structures or installations. Soil compaction is a repetitive process and much can be gained from properly planned and executed compaction trials [14, 15].

26.1.3 GPS and GSM

Atomic clocks are currently the most precise instruments known, losing a maximum of 1 s, every 30,000–1,000,000 years. In order to make even more accurate, they are regularly adjusted or synchronized from various control points on Earth.

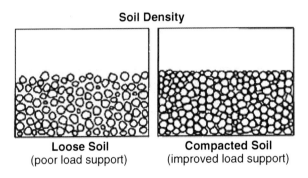

Fig. 26.1 Soil compaction[4]

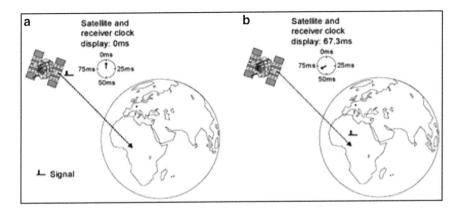

Fig. 26.2 (a) Signal transmission. (b) Signal reception

Each satellite transmits its exact positions and its precise on board clock time to Earth at a frequency of 1,575.42 MHz's [6].

Figure 26.2 represents about the communication between receiver and the satellite. If the signal is transmitted from the receiver at 0 ms, for a normal GPS receiver should have atleast 67.3 ms to receive the signal from the satellite.

The architecture of GSM network is represented in Fig. 26.3. The GSM architecture can be classified into four main parts they are Mobile Station (MS), Base Station Subsystem (BSS), Network and Switching Subsystem (NSS), and Operation and Support Subsystem (OSS). Short Message Service (SMS) was a very popular service offered by GSM operators. The SMS is a kind of paging with acknowledgement of message delivery. The Global System for Mobile communication is comes under digital cellular communication system for mobile users. GSM was designed to be compatible with ISDN services. The technical specification for the Global System for Mobile communication has been produced by the Special Mobile Group (SMG) Technical Committee (TC) of the European Telecommunications Standards Institute (ETSI) [2].

Fig. 26.3 GSM network[2]

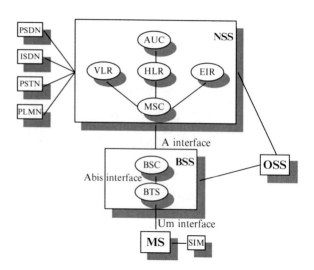

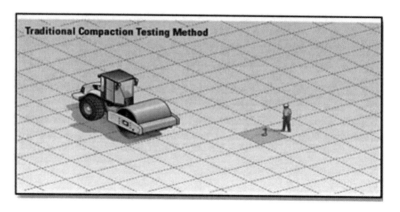

Fig. 26.4 Compaction testing method[1]

26.1.4 Integrating GPS with Compaction Measurement

Mapping refers to the positioning technique which makes maps the current location in global mapping resources like Google, Gmap etc. (Fig. 26.4)

The proposed cost-effective approach will address

- Precise location monitoring of compaction machine
- Gives feedback to the driver about the compaction level
- Reducing the delay of determining accurate position of vehicle
- Determines the reputation occurred by various compactors
- Real time updates between the driver and server room by GSM/GPRS technology

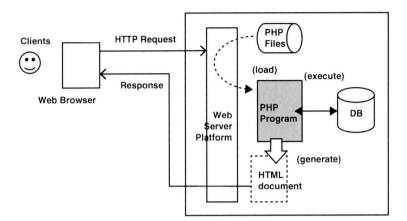

Fig. 26.5 Remote web server[12]

26.1.5 Remote Web Server

The main purpose of tracking server is to monitor and maintain all information receiving from Vehicles into a central database. This database is accessible over internet to valid users through a web interface and the valid users can also track their vehicle and view all previous information stored in database. This saves the information into database. The various fleet levels monitored namely Vehicle Location, Total fuel used, Fuel level (0–100 %), Total engine hours (h), Total Working Hours, Idle Time and Soil Compaction level.

User interface design for the application is designed using Hypertext mark-up language, Java Script and Cascading Style sheet where the application has become a user interactive and satisfies user needs. Server side validations are done using PHP (Hypertext Pre-processor) to validate the server side actions. Web applications commonly use a combination of server-side script and client-side script for Communicating with the users. The client-side script deals with the presentation of the information as front end to the end users while the server-side script deals with all the hard stuff like storing and retrieving the information. Both are interfaced using XAMMP Server.

The corresponding file named test.php is loaded and put into execution. The application program may access persistent data stored in MySql database and generate an HTML document as a http response. The generated HTML document appears on the Web browser for the client to manipulate as shown in the Fig. 26.5.

26.2 Related Work

Jinsang Hwang, Hongsik Yun in title Development of Soil Compaction Analysis Software (SCAN) Integrating a Low Cost GPS Receiver and Compactometer [7] has design software for soil compaction analysis (SCAN). The SCAN is distinguished from other previous software for intelligent compaction (IC) in that it can use the results from various types of GPS positioning methods, and it has an optimal structure for remote managing the large amounts of data gathered from numerous rollers. Abid Khan, Ravi Mishra in the paper GPS – GSM Based Tracking System [8] develop a tracking unit that uses the global positioning system to determine the precise location of a object, person and other asset to which it is attached and using GSM (Global System For Mobile) modem this information can be transmit to remote user. The design is cost-effective, reliable and has the function of accurate tracking. Automobile localization system using GSM/GPRS transmission was implemented by Ioan Lita et al. [11]. In this paper the vehicle position is tracked by the GPS module and the position of the vehicle is transmitted to the owner as SMS. The GPS and GSM module are interfaced using a PIC microcontroller.Bharat Kulkarni implemented a GSM based automatic meter reader system. In this paper the user can collect the electric bill in his phone through GSM. The interfacing between GSM using ARM processor [10]. Vehicle tracking using GPS was implemented by T. Krishna Kishore et al. and in this work they present the principles of a low operational cost flexible internet based data acquisition system [9].

26.3 Proposed Method

The proposed implementation on Location based Compaction Level Monitoring System focus on over coming unbalanced compaction terms in corresponding location for a soil compactor with remote monitoring system. The proposed work involves the following modules (1) Configuring a GPS receiver for XYZ co-ordinates to be obtained from NMEA data. (2) Integrating GPS data with Soil Compaction Measurement unit. (3) Sending both the data to the remote server and the alert messages to mobile phone via GSM modem [13]. (4) Integrating the GSM data in the user interactive Web based application for the end users. (5) Designing GUI at driver level to represent with advantages like Compaction Level Monitoring, Graphical representation, Real time mapping in GMap, Engine Status etc.

The functional diagram for the proposed system is given by (Fig. 26.6). GPS and compaction data will be integrated and given to the processor. Then the two data may send to the GSM modem. From that the data will transmitted to the remote server. In the remote server all the hard stuffs are happening. With the help of the another GSM receiver modem the data is received in the remote server through command prompt which is further stored in the database for further process (Fig. 26.13). For the data stored in database the location and the compaction details are viewed through the user interacted web based application.

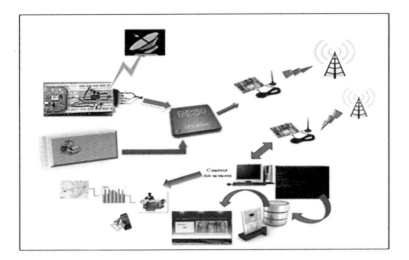

Fig. 26.6 Functional diagram

26.3.1 GPS Data Acquisition

Global Positioning System satellites provide precise location information for slope and elevation control with centimetre-level accuracy [10]. Each NMEA data has its own data type. Each data type has its interpretation sentences. The receiving unit just checks the checksum and ignores if the checksum is bad. This data from the respective receiver is interfaced with the ARM7/TDMI processor. This gives latitude and longitude of the vehicle location. In the NMEA standard there are no commands to indicate that the GPS should do something. Instead each receiver just sends all the data and expects much of it to be ignored. Some receivers may have commands inside the unit that can select a subset of all the sentences.

26.3.2 Hardware Description

The project uses ARM7 family LPC 2378 processor for interfacing the GPS and the GSM module. The ARM7/TDMI is a general purpose 32-bit microprocessor which offers high performance and very low power consumption. GPS interfaced with the processor to collect the location data. The collected data may be transmitted to the remote server through GSM which will be interfaced with the processor. Location data and the compaction data will be sent to the processor through serial

Fig. 26.7 Raw compacted reading[1]

Volume of Mold $n(ft^3)$	Weight of wet soil in the mold (lb)	Wet Unit Weight (lb/ft^3)	Water Content (%)	Dry Unit Weight (lb/ft^3)
1/30	3.88	116.4	12	103.9
1/30	4.09	122.7	14	107.6
1/30	4.23	126.9	16	109.4
1/30	4.28	128.4	18	108.8
1/30	4.24	127.2	20	106.0
1/30	4.19	125.7	22	103.0

communication. For serial communication [3], some registers in the processor has been enabled. Enabling the control functions of the pins the pin select register has been triggered and the line control register is to set the baud rate for the serial transmission. The data is queued in the transmit hold register this data is sent via GSM and in order to receive the data from the GPS and compaction, receive buffer register should be triggered.

26.3.3 Compaction Monitoring

Compaction area is divided into various field of yard. Compaction data is obtained from the compactometer which is attached in the vehicle. Method of computing the compacted data with the current location play a vital role in the system.

Figure 26.7 shows a raw compacted reading from a standard proctor test [5]. This shows the number of pass by the vehicle, wet unit area and the dry units are taken for computing the compaction process. This depends upon the water content of the soil in current location.

Figure 26.8 tabulates about the compaction level of various compactor in their location. Minimum value of compaction is represented by LP1 low and maximum value of compaction is represented LP2 high. Assumed values for both LP1 low and LP2 high are represented above.

Simulated Compaction P1, P2, P3 is checked with a threshold condition and status of the threshold is obtained (Fig. 26.9). Corresponding message may send by the processor ARM7/TDMI according to the conditions.

Threshold has been checked with respect to the current location (Fig. 26.10). Compaction level will calculate from one of the processor peripheral and value is sent to the communication devices.

Fig. 26.8 Compaction threshold for corresponding location

LOCATIONS OF VARIOUS SOIL COMPACTORS IN THE YARD	COMPACTION LEVEL
L_1	$LP_{1\,low}$ - $LP_{1\,high}$
L_2	$LP_{2\,low}$ - $LP_{2\,high}$
L_3	$LP_{3\,low}$ - $LP_{3\,high}$

Assume: $LP_{1\,low}$-800 $LP_{2\,low}$-700 $LP_{3\,low}$-900
$LP_{1\,high}$-850 $LP_{2\,high}$-750 $LP_{3\,high}$-950

SIMULATED COMPACTION	THRESHOLD CONDITION	PROCESSOR OUTPUT	THRESHOLD STATUS	MESSAGE
P_1	$LP_{1\,low} < P_1 < LP_{1\,high}$	820	YES	Threshold attained
P_2	$LP_{1\,low} < P_2 < LP_{1\,high}$	680	NO	Threshold not attained
P_3	$LP_{1\,low} < P_3 < LP_{1\,high}$	940	YES	Threshold attained

Fig. 26.9 Comparison of processor output with threshold conditions

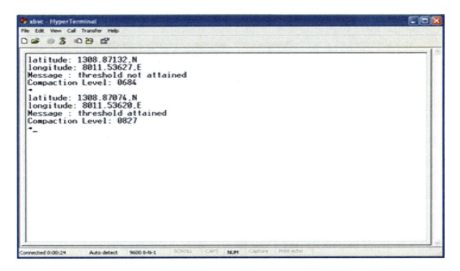

Fig. 26.10 Threshold checking with GPS data and corresponding compaction level

26.4 Result

The communication of the GPS data and the compaction data to the mobile interface was tested by a Samsung mobile phone.

Figure 26.11, shows the Hyper Terminal output for sending the acquired data. Required AT Commands was generated by the processor. GPS data with corresponding compaction level has been checked and sent to a GSM network.

Figure 26.12 shows the alert messages that are sent to the mobile phone for various compaction values in the specified location. The message shows the latitude and longitude value of the vehicle and also the corresponding compaction value for the location were also computing by the processor (Fig. 26.13).

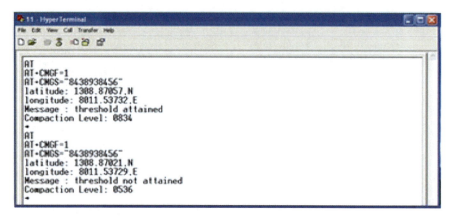

Fig. 26.11 HyperTerminal output for sending data

Fig. 26.12 Data received through GSM

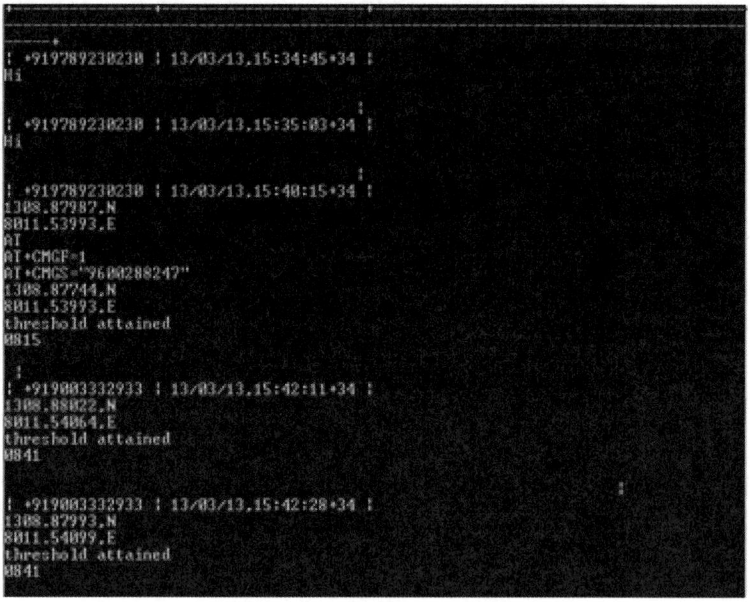

Fig. 26.13 GSM data received through command prompt in server

26.5 Conclusion and Future Work

Thus in this paper we provide better accuracy with low cost GPS receiver's was completed successfully. GPS Co-ordinates with compaction level has been acquired successfully. A simulated set of required compaction data has been obtained from the controller and the thresholds were verified by mapping with location data. Remote monitoring was attained by the GSM technology.

As an extension of this work a GUI will be designed. GUI will contain a view of Gmap which shows the vehicle's current position in the yard. Compaction measuring table provides different color codes for comfort of the driver to understand the location based compaction level. The data will be sending to the remote server using GPRS communication for continuous transmission of the data.

References

1. Accugrade Compaction, *GPS Mapping and Measurement for Soil Compactors*, (2007)
2. GSM Technical Specification, *GSM 11.11*, Dec 1995, version 5.0.0
3. GSM Technical Specification, Vi Microsystems Pvt. Ltd, Rev. 4, 26 July 2012
4. http://www.en.wikipedia.org/wiki/Soilcompaction

5. Soil Mechanics & Foundation Lecture 3.3 pp. 10, Available at http://www.engr.uconn.edu/lanbo/CE240LectW033Compaction.pdf
6. http://share.pdfonline.com/2e0cdab915644f288e3277548769e136/Introduction20to20GPS.html
7. J. Hwang, H. Yun, J. Kim, Y. Suh, S. Hong, D. Lee, *Development of Soil Compaction Analysis Software (SCAN) Integrating a Low Cost GPS Receiver and Compactometer*. Available at www.mdpi.com/journal/sensors, ISSN 1424-8220, Sensor 2012, pp. 2352–2353
8. A. Khan, R. Mishra, GPS – GSM based tracking system. Int. J. Eng. Trends Technol. **3**(2) (2012)
9. T. Krishna Kishore et al., Vehicle tracking using a reliable embedded data acquisition system with GPS and GSM. Int. J. Comput. Sci. Netw. Secur. (IJCSNS) **2** (2010)
10. B. Kulkarni, GSM based automatic meter reading system using ARM controller. Int. J. Emerg. Technol. Adv. Eng. **2** (2012)
11. I. Lita et al., *A New Approach of Automobile Localization System Using GPS and GSM/GPRS Transmission*, 1-4244-0551-3/06 ©2006 IEEE
12. S. Nakajima, *Architecture of Dynamically Adaptive PHP-Based Web Applications*, ISBN:1530-1362/11 ©2011 IEEE
13. R. Prakash, K. Suresh, B. Venkatalakshmi, S. Vijayakumar, Development of pervasive compaction monitoring interface for soil compactor – a GPS/GSM based approach, in *International Conference on Recent Trends in Information Technology*, MIT, Anna University, Chennai, India, ISBN:978-1-4799-1024-3/13 ©2013 IEEE
14. M.K. Rainer, *TransVib 2002 – International Conference on Vibratory Pile Driving and Deep Soil Compaction*, Louvain-la-Neuve. Keynote Lecture (2002), pp. 33–42
15. D.J. White, Field evaluation of compaction monitoring technology. Partnership for Geotechnical Advancement, Iowa State University, Mar 2006, pp. 1–4. Available at www.mdpi.com/journal/sensors

Author Index

A
Ahmed, P., 195
Ambika, D., 171
Amrita, A., 195

B
Bennur, V.S., 139
Bhattacharya, S., 21
Bhol, M., 45
Brahmkstri, K., 265

C
Chaukwale, R., 81
Chowdhury, A., 275
Chowdhury, F., 3

D
Dhinakaran, B.C., 229

F
Furness, J., 3

G
Geetha, G., 213
Geetha, V., 21

H
Hariharan, K., 297
Helmy, T., 93
Howlader, J., 275

I
Iyengar, S.S., 31

J
Jadhav, A., 265
Jahan, T., 161
Jariwala, V., 125
Jayashri, S., 149
Jinwala, D.C., 125
Joshi, R.C., 183

K
Kamath S.S., 81
Kaur, K., 213
Kaviya, P., 307
Kolberg, M., 3
Koushik, C.P., 319
Kshirsagar, D.D., 265
Kumar, P., 125
Kumaraswamy, M., 31

L
Lee, J.-K., 229

M
Mc Gonigle, S., 255
Meghanathan, N., 55
Mishra, P., 183
Mishra, S.K., 45
Mohanty, M.N., 45
Mydhile Shanmugam, S., 319

N
Nagamalai, D., 229
Narasimha, G., 161
Nath, S., 275
Neelamegam, A., 243
Nuallain, E.O., 255

O
Okatan, A., 229
Ozcan, A., 229

P
Patnaik, L.M., 31
Paulraj, P., 243
Pilli, E.S., 183
Prakash, R., 319

R
Radha, V., 171
Raj, T., 21
Ramalakshmi, R., 307
Ramaprasath, A., 297
Ranjan, P., 183
Rao, C.V.G., 161
Rao, G.R.K., 281
Rawat, J.S., 183

S
Sastri, A.P., 281
Sawant, S.T., 265
Shaila, K., 31
Shirabur, S.S., 139
Shukla, A., 107
Singh, N., 69
Sreenivas, R.S., 69
Sridhar, 21
Srinivasan, A., 297
Subha, T., 149
Sur, C., 107
Suresh, K., 319
Sutagundar, A.V., 139

T
Taylor, A., 255
Tejaswi, V., 31
Thomas, D., 265

V
Venugopal, K.R., 31
Vutukuri, A., 21

W
Wang, M., 255
Wang, Q., 255

Subject Index

A
ACO. *See* Ant colony optimization (ACO)
Active node selection, 311–312
Active sensors, 308–310, 314, 315, 317
Ant colony optimization (ACO), 81–91, 108, 113, 114, 119
ARM7/TDMI, 319, 325, 326
Auditability, 155

B
Bandwidth optimization, 299, 300, 303, 304
Bit error rate (BER), 46–53
Bloom filter, 125–136
Business intelligence, 243–253

C
CDS. *See* Connected dominating set (CDS)
CFS. *See* Correlation-based feature selection (CFS)
Channel capacity, 46, 51
Cipher text, 187
Classification, 33, 162–167, 169, 177–179, 187, 188, 196, 197, 201, 202, 210, 245, 268
Clear channel assessments (CCAs), 22, 23
Cloud computing, 81, 150, 151, 184–187, 195, 229, 292
Cloud service provider (CSP), 151, 152, 183–186, 188–190, 193
Clustering, 162–166, 169, 178, 179, 243–253, 268
Cognitive radio network (CRN), 256–259, 261, 263, 264
Combinatorial optimization, 82, 108, 120
Community-based network, 258
Computational trust, 257
Connected dominating set (CDS), 56, 307–317
Consistency-based feature selection (CON), 197, 199–201, 206–210
Correlation-based feature selection (CFS), 197, 199, 201, 206, 209, 210
CRN. *See* Cognitive radio network (CRN)
CSMA-CA mechanism, 22–25, 28, 29
CSP. *See* Cloud service provider (CSP)

D
Data distortion, 162–164, 166
Data gathering, 55–68
Data mining (DM), 162–166, 169, 243, 244, 267
DDoS, 230–238, 240, 241, 282
Decision making, 112, 119, 244, 246, 253, 266, 268
Decryption, 186–188, 276
DEFLATE algorithm, 301
Denial of service (DoS), 184, 198, 210, 234, 236, 237, 266, 267, 271, 281–292
Digital signature, 125–136, 287
Discrete-time queues, 69–79
Distributed hash tables (DHTs), 3
Distributed IDS, 266, 267

E
E-commerce, 94, 244, 245, 250, 251
Egyptian vulture optimization algorithm, 107–120
Eigen value, 50, 52, 53
ELGamal, 186, 188–190, 192, 193
Elliptic curve digital signature algorithm (ECDSA), 125–136

Email phishing, 238–239, 241
Email security threats, 229–241
Encryption, 150, 151, 184, 186–190, 192, 193, 276, 304
EpiChord, 3–18
Event detection and reporting algorithm (EDRA), 4, 5

F
Factorization, 273, 276, 279
Feature extraction, 174–175, 182
Feature selection, 162, 195–210, 268
Frequent itemset, 243–253
Fuzzy logic, 161–169
Fuzzy membership functions, 162, 164, 166, 169

G
Gain ratio (GR), 197, 199, 201, 202, 206, 209, 210
GIT repository, 300, 304
GIT subversion control system, 299
Global positioning system (GPS), 140, 143, 320–329
Global system for mobile (GSM), 320–326, 328, 329
GPRS, 301, 302, 322, 324, 329
GPS. *See* Global positioning system (GPS)
Graph based problems, 108, 112, 120
GSM. *See* Global system for mobile (GSM)
GUI design, 225

I
IaaS, 150, 157, 183
IDS. *See* Intrusion detection system (IDS)
IEEE 802.15.4 WSN Standard, 21–29
Information gain (IG), 196, 197, 199–202, 206, 209, 210, 241
Integrity, 126, 136, 149–158, 187, 188
Intrusion detection system (IDS), 99, 195–210, 234, 236, 237, 265–273

J
Job shop scheduling, 81–91

K
k-means clustering, 165, 166, 178, 243–253
KDD cup 99 dataset, 196–199, 202, 206, 267, 270
Key exchange, 275–280

L
Load balancing, 82, 86, 88–91
Localization, 139–145

M
MATLAB, 166, 225
Maximum stability, 55–68
Medium access control, 21, 31–42, 60, 69
Mel frequency cepstral coefficients (MFCC), 171–182
MIMO. *See* Multiple-input multiple-output (MIMO)
Minimum distance spanning tree, 55–68
Mobile networks, 18, 34, 108, 297–304
Mobile sensor networks, 55–68
Modern network architectures, 232
Modulo, 277
Multi-agent system (MAS), 120, 265–273
Multiple-input multiple-output (MIMO), 46–51, 53

N
Network lifetime, 40, 56, 57, 61–68, 143, 145, 308–310, 317
Network Simulator 2 (NS2), 8, 25, 38, 70, 76
Nigerian scam, 239–241
Node lifetime, 56, 61, 64–68

O
Ontology, 265–273
OverSim, 3–18

P
P2PSim, 7, 8
Packet delivery ratio, 78, 79, 314–316
Packet spoofing, 282, 288–289
Peer to peer system, 8
Phishing, 230, 231, 234, 235, 237–241
Plain text, 186, 191
PlanetSim, 8
Prime numbers, 277
Private key, 126–129, 132, 154, 186, 188–190, 258
Public key, 127, 128, 132, 154, 156, 186, 188, 189, 257, 277

Subject Index

Q
Quality of service (QoS), 22, 23, 32, 33, 35–38, 41, 42, 70, 183
Queue back-pressure random access algorithms (QBRA), 70

R
Radio environment mapping (REM), 255–257, 259, 261, 262
Random access protocol, 32, 33, 41
Recommender system, 245, 251
REM. *See* Radio environment mapping (REM)
Remote procedure calls (RPC), 8, 9
RNA-FINNT, 213–228

S
Scalability, 4, 7, 9, 17–18, 150, 153, 184, 245
Self learning, 265–274
Sensor scheduling mechanism, 308–310, 317
Sensory manipulation attacks, 257
Shortest path, 83
Smurf attacks, 273, 282, 287, 288, 290–292
Soil compaction measurement, 324
Spam, 230–241
Spatial correlation, 46–49, 53
Speaker modelling, 177
Speaker recognition, 172–174, 182
Spearhead phishing, 239
ST-MAC protocol, 76–78
Storage expansion, 149
SYN flooding, 282–285

T
Task scheduling, 81–91
TCP hijacking, 282, 286–287, 290
Time reversal (TR), 47, 49–53, 100
Trilateration, 141–143, 145
Trust management systems, 258
Trust model, 94–95, 103, 257, 259
Trusted third party auditor (TTPA), 149–158

U
Ultrawideband (UWB), 45–47, 49, 50

V
Vector quantization, 165, 171–182
Vehicle routing optimization, 82

W
Web services, 150, 153, 157, 265–273
Wireless networks, 22, 34, 69–79, 309
Wireless sensor network (WSN), 21–29, 31–42, 125–136, 139–146, 307–317